Bat Roosts in Trees

A Guide to Identification and Assessment for Tree-Care and Ecology Professionals

Bat Tree Habitat Key

Pelagic Publishing | www.pelagicpublishing.com

Published by
Pelagic Publishing
PO Box 874
Exeter
EX1 9YH
www.pelagicpublishing.com

Bat Roosts in Trees
A Guide to Identification and Assessment for Tree-Care and Ecology Professionals

ISBN 978-1-78427-161-9 *Paperback*
ISBN 978-1-78427-162-6 *ePub*
ISBN 978-1-78427-164-0 *PDF*

British Library Cataloguing in Publication Data
A catalogue record for this book is available from the British Library

Cover images: The treasures that may be discovered by applying the methods described in this text include: male noctules *Nyctalus noctula* 'singing' from woodpecker-holes in the autumn mating season (photo: Jasja Dekker); colonies of whiskered bats *Myotis mystacinus* tucked behind bark in the spring (photo: René Janssen); and, barbastelles *Barbastella barbastellus* hiding in plain sight at any point in the year (photo: Henry Andrews).

Contents

Preface

The objective that was set at the outset of this book was to produce a set of instructions that would enable anyone with an interest in tree-roosting bats, anywhere in the British Isles, to make an objective assessment of any wooded habitat type for its potential to hold bat roosts.

In order that the instructions were accessible, they were to be in plain English and following a logical progression.

The finished article has a bias to professional surveys. However, in the final analysis, it is concluded that the objective has been achieved.

Nevertheless, the reader should keep in mind that the variables that influence any aspect of the natural world are like a multitude of spheres, each of which is densest in the centre. These spheres represent the individual bat, the age of the bat, the sex of the bat, the species of the bat, the prevailing weather, the climatic trend, the season, the individual tree, the tree species, the habitat, the environment … the complexities are endless! Without any pressure bearing on them, the spheres are perfect orbs, but the spheres are never static and as they move they are pushed and pulled in a multitude of ways, sometimes with wide spaces between, and sometimes squashed together. In some combinations, the spheres may melt into each other like the bulbs of wax in a lava-lamp, only to separate entirely again and then repel each other like opposing magnetic poles when the combination is even so much as subtly different.

The scope of this book is such that it must rely to a degree on generalisations. These generalisations represent the firm centre of the individual spheres that are the least subject to being squashed or stretched: the solid ground. The further away from the centre, the more uncertain the situation becomes, and these uncertainties cannot be shoe-horned into a field-guide. The longer the book exists, the more people will encounter situations that confound the generalisations.

In order to anticipate this situation, the book exists as a framework that relies on the living cells of the BTHK Database. It is hoped that the use of the book and the Database will result in more roosts being found, more records being submitted, and the strength of the analysis increasing to a point where generalisations are no longer required.

Until that day, the reader should keep in mind the words of Douglas Bader: *'Rules are for the obedience of fools and the guidance of wise men.'* Some of the *evidence* reviewed is unequivocal and to ignore it would be irresponsible. The *advice* is experience-based and should not be applied as a set of rules without taking into account Bader's maxim. It is nothing more than an account of methods that have been found to be broadly effective. Although the methods might be adapted by an individual to suit his or her own project, they may equally be abandoned entirely in favour of something else that better suits the talent and skill of the individual.

Now read on …

Citing, Credits and Acknowledgements

Citing

Where reference is made to any part of the content of this book, please cite as 'BTHK (2018)', giving the full reference as:

» **BTHK 2018.** *Bat Roosts in Trees – A Guide to Identification and Assessment for Tree-Care and Ecology Professionals.* **Exeter: Pelagic Publishing.**

Bat Tree Habitat Key is a registered trademark.

Credits

Text

Chapter credits are as follows:

» Chapter 1 – Henry Andrews.
» Chapter 2 – Henry Andrews.
» Chapter 3 – Henry Andrews.
» Chapter 4 – Henry Andrews.
» Chapter 5 – Henry Andrews.
» Chapter 6 – Henry Andrews.
» Chapter 7 – Henry Andrews.
» Chapter 8 – Henry Andrews and Mark Gardener.
» Chapter 9 – Henry Andrews and Mark Gardener.
» Chapter 10 – Henry Andrews.
» Chapter 11 – Henry Andrews (with critical review by Jim Mullholland).
» Chapter 12 – Henry Andrews.

Henry Andrews

Henry Andrews is a chartered ecologist and coordinator of the *Bat Tree Habitat Key* project. He has a particular interest in structure-based predictive assessments and the use of detailed biological data within structured frameworks to predict species occurrence. He owns a small consultancy practice in the south-west of England, specialising solely in the minerals industry. Email: henry.andrews@aecol.co.uk

Mark Gardener

Mark Gardener is an ecologist and lecturer. He is the author of several books on ecology and data science (including statistical programming) and is responsible for the statistical analysis used in this book, and all scientific investigation performed by the BTHK project. He currently teaches for the Open University as well as running freelance courses in ecology and data science. Email: mark.gardener@me.com

Photographs

Chapter 3 – PRFs

BTHK is grateful to the following for providing individual roost examples used in the PRF figures: **Paul Kennedy** for showing me the barbastelle roost in the lightning-strike (top row), the **Wytham Bat Project** for finding the Natterer's roost in the canker; **Daniel Whitby** for showing me the Natterer's bat roost in the beech compression-fork (top row); **Garry Mortimer** for finding the Natterer's bat roosts in the Scots pine compression-fork and **Keith Cohen** for providing the photographs (bottom row); **James Shipman** and **Sam Davis** for showing me the lesser horseshoe-bat roost in the Turkey oak butt-rot and providing the image of the bats (top row), **Stewart Rowden**, **Jenny Greenwood** and **Alasdair Grant** for finding the lesser horseshoe-bat roost in the London plane (middle row) and **Iain Hysom** for showing it to me; **Daniel Bennett** and **Sarah Allen** for finding the lesser horseshoe-bat roost in the lime (bottom row) and **Richard Crompton** for telling me how to find it; **Steve Allen** for finding the Natterer's bat roost in the shearing crack and showing it to me (top row); **Jon Russ** for providing the photographs of the barbastelle roost in the weld (top row); **Hannah Montag** for showing me the noctule roost in the lime weld (middle row); **Hannah Haggon** and **David White** for showing me the brown long-eared bat down behind bark (bottom row); **Chris Koczy** for finding the brown long-eared bat in the beech flute and **Richard Koczy** for providing photographs (middle row); **BSG Ecology** for sharing the common pipistrelle ivy record made using a thermal imaging camera.

Chapter 8 – Close-Inspection

BTHK is grateful to the following for providing individual field-sign examples: **Jim Mullholland** for the images of Bechstein's bat-flies (Figure 8.7); and **Carla Bosworth**, **Chris Watts** and **Matt Dodds** for finding the Daubenton's bat roost in the wet PRF (Figure 8.14). BTHK is also grateful to **James McGill** for the invertebrate and bird identification.

Acknowledgements

The *Bat Tree Habitat Key* project was born from modules taught by **Sarah Wild** and **Alex Lockton** during the MSc in Biological Recording (now hosted by Manchester Metropolitan University). Their teaching lead to the 'Eureka moment' of realisation that the greatest value of biological records was their predictive power. By investigating the idea during assignments, it became apparent that these data were the alchemic stardust that could be mixed to define the algorithm that would render the invisible visible, not just when the target was hiding within a habitat resource, but also when the target was absent altogether. Without Sarah and Alex's inspiring talks in the classroom and the bar, the project would never have been imagined. *Bat Tree Habitat Key* therefore owes them its life!

This book would not have been possible if it were not for lengthy conversations with **Danielle Linton** (Wytham Bat Project), **Daniel Whitby** (AEWC Ltd) and **Erik Korsten** (Bureau Waardenburg). The *Bat Tree Habitat Key* project owes them immeasurable gratitude for their experience, wisdom and candour.

The practical knowledge of the tree-roosting ecology of whiskered bats, Brandt's bats and Nathusius' pipistrelle comes from the generosity of **René Janssen** (Bionet Natuur-onderzoek) and **Mark Van De Sijpe** (Bat Group Natuurpunt, Belgium), and that of Leisler's bat from **Austin Hopkirk** and **John Russ** (Hopkirk & Russ Bat Ecology).

The accounts of ivy roosting would have been incomplete were it not for the photographic evidence provided by **BSG Ecology**, and the account of fluting would have been non-existent were it not for the massive body of data provided by **Ben McLean**.

The power of the analysis was significantly increased by the contribution of a huge body of data from the Tortworth Forest Estate by **Jim Mulholland** (Mulholland Ecology & Arboriculture).

Our understanding of just how important sycamore is as a roost tree was only possible following the discovery of three small woodlands holding a superabundance of wounded trees by **Carrie White**. Since then, over 100 arborists and ecologists have significantly improved their knowledge of tree-roosting ecology looking in roosts Carrie first recorded.

Finally, the project could not continue without the support of **Theresa Radcliffe, Erin Andrews, Eugene Andrews, Isobelle Andrews** and **Tristan Andrews**, who have endured long walks over uneven terrain and deep mud (which are more uneven and much deeper when you're only two-foot-tall), yet always in good humour and with great determination.

CHAPTER 1

Rationale

In this chapter	
Introduction	Curiosity and necessity
Pre-existing published advice	Its effective application
Motivation	A cursory review of legislation
Objective	Policy and the threshold of *"reasonable likelihood"* ('more-likely-than-not')
Proportionality	Balance; ensuring the ends justify the means

1.1 Introduction

There is more than one reason for reading this book, but all reasons can be divided into two broad camps:

» Curiosity.

» Necessity.

If you have opened this book because you are curious, then you know your level of motivation and what you hope to gain from the exercise.

If you have opened this book out of necessity, it is possible you do not want to read it and may not be motivated at all.

The irony is that those of you reading out of curiosity may read this chapter or not as you choose, but those of you reading out of necessity, even if you have little or no motivation, *must* read this chapter.

Let there be no misunderstanding: every tree is contentious and a woodland is an anvil waiting to fall *Looney-Tunes*-style upon the career of the unwary.

There is *always* public opposition to any operation that may fell trees. If you are assessing an area of woodland in support of a development proposal, you may well find yourself in the local paper. If you are assessing Ancient Semi-Natural Woodland in support of a development proposal, it is well within the bounds of probability that you will end up facing a Public Inquiry and national infamy, so you would be wise to ensure your appraisal is robust, because the opposition may bring in a hired-gun to try and shoot it full of holes.

In the recent past it has been common practice for appraisals of wooded habitat to defer to published guidance, in an (entirely understandable) attempt to guard against the possibility that a client might fall foul of conservation legislation, and to avoid any challenge from a third-party (such as a Local Authority Ecologist).

1.2 **Pre-existing published advice**

At the time of writing (2018), there are currently two publications that deal with roost surveys in wooded habitat:

» British Standards Institute 2015. *BS 8956 – Surveying for bats in trees and woodland.* BSI London; and

» Collins J. (ed.) 2016. *Bat Surveys for Professional Ecologists – Good Practice Guidelines.* London: Bat Conservation Trust.

Both include sensible advice and both have an entirely honourable foundation. Nevertheless, this book will not refer to either of these guidance documents again following this chapter. The reason for this is that routine deference to published guidance (without having reviewed the differences between each individual species, and thereon tailored the actions advocated to the most effective equipment, method, timing and effort in the context of a specific project) dulls the edge of the surveyor. To a certain extent, it also denies the surveyor the satisfaction of designing the survey, and ultimately the joy of performing it.

Thankfully, the authors of both guidance texts were sufficiently experienced as to anticipate this, and both allow for creative input in the design of appraisals.

The British Standard opens with this statement:

"As a guide, this British Standard takes the form of guidance and recommendations. It should not be quoted as if it were a specification or a code of practice and claims of compliance cannot be made to it."

This is also the spirit of text within subsection 1.1.3 of Bat Conservation Trust's *Good Practice Guidelines*, which states:

"The guidelines should be interpreted and adapted on a case-by-case basis according to site-specific factors and the professional judgement of an experienced ecologist. Where examples are used in the guidelines they are descriptive rather than prescriptive."

Furthermore:

"It is accepted that departures from the guidelines (e.g. either increasing or decreasing the number of surveys carried out or using alternative methods) are often appropriate. However, in this scenario an ecologist should provide documentary evidence of (a) their expertise in making this judgement and (b) the ecological rationale behind the judgement."

In this context 'descriptive rather than prescriptive' might be taken to mean that the examples are not rules that must be enforced. Nevertheless, although the guidance is not a book of rules, any deviation should be supported by tangible evidence that demonstrates unequivocally the following criteria:

» The circumstances warranted the deviation.

» The method and intensity employed can be proven to be appropriate for the circumstances.

In essence, the requirement is for evidence-supported action, which is both reasonable and sensible. An approach that has been found to satisfy this requirement has been to design the survey and, when the team is satisfied that the design will collect robust data that is justifiably necessary and that can be meaningfully interpreted within a repeatable framework, to compare their design with BS8956 and the guidance produced by Collins (2016) to see how far their design deviates, and why. This ensures the recommendations are supported by evidence to which the reader of any subsequent report will have access, and provides the reader with sufficient information to allow them to perform an independent critical appraisal of the rationale adopted. The process is broadly this:

Step 1   Review the legislation, planning policy and case-law, and any pertinent consultation in order to define an interpretation threshold against which data may be compared.

Step 2   Collate the existing scientific evidence and ensure copies of any texts that were referred to in the survey design are available for third-party review.

Step 3   Gather and collate pre-existing intelligence relating to the site.

Step 4   Review the existing intelligence and conduct a proportionality test to decide whether surveillance is appropriate.

Step 5   If surveillance is appropriate, review the methods available and chose the most effective suite to suit the context of the site and operation proposed (i.e development, management action, etc.).

Step 6   Identify constraints; acknowledge them and mitigate where possible.

Step 7   Define the analysis framework and identify the predicted outcome.

Step 8   Perform the survey.

Step 9   Interpret the results within a repeatable framework.

Step 10   Acknowledge any failings and, where possible, make suggestions as to how these might be overcome.

Step 11   Actively encourage critical attack in order that any unidentified failings that may potentially exist are found and given due attention.

1.3 Motivation

1.3.1 The law

For most tree-care professionals, ecological surveyors and consultants, our motivation to search for bat roosts is founded in the law, and our desire to remain on the right side of it. There are currently two mechanisms that spur our action:

» The *Wildlife & Countryside Act 1981 (& as amended).*

» The *Conservation of Habitats and Species Regulations 2017.*

The two pieces of legislation are set out below.

Wildlife & Countryside Act 1981

All bat species are listed under Schedule 5 of the *Wildlife & Countryside Act* 1981 and receive legal protection under Part 1, section 9, sub-section (4)(b) and (c) which states:

> *"Subject to the provisions of this Part, a person is guilty of an offence if intentionally or recklessly—*
> *(b) he disturbs any such animal while it is occupying a structure or place which it uses for shelter or protection; or*
> *(c) he obstructs access to any structure or place which any such animal uses for shelter or protection."*

Conservation of Habitats and Species Regulations 2017

All bat species are listed under Schedule 2 of the *Conservation of Habitats and Species Regulations 2017.*

Part 3, regulation 41, paragraph (1) of the *Conservation of Habitats and Species Regulations 2017* states:

"A person who —

(a) *deliberately captures, injures or kills any wild animal of a European protected species,*
(b) *deliberately disturbs wild animals of any such species,*
(c) *deliberately takes or destroys the eggs of such an animal, or*
(d) *damages or destroys a breeding site or resting place of such an animal,*

is guilty of an offence."

> **Note:** The offence of damaging or destroying a breeding site or resting place does not include the word 'deliberately', but is an *absolute* offence that does not require any fault elements to be proved to establish guilt.

Part 3, regulation 41, paragraph (2) states that disturbance of animals includes any disturbance which is likely:

"(a) to impair their ability —
    (i) *to survive, to breed or reproduce, or to rear or nurture their young, or*
    (ii) *in the case of animals of a hibernating or migratory species, to hibernate or migrate;*
      or
(b) *to affect significantly the local distribution or abundance of the species to which they belong."*

1.4 Objective

Although the motivation for performing a survey may have a legislative foundation, the discipline is not an exact science – the law cannot therefore be easily used to define a robust objective, and without a clear objective the interpretation of surveillance results will be handicapped. To explain: the motivation is why the appraisal is performed, and the objective is what the appraisal aims to achieve.[1]

The fundamental difference between good and bad appraisals is the ability to define the objective.

A good objective is straightforward and single-minded. It may only comprise a component part of an overall wider campaign, but it is palpable; you can see it clearly and hold it in your mind. A good objective ensures the appraisal is not distracted or side-tracked because it knows its weight, worth, and why it is needed, it can therefore justify its own existence and stand its ground when challenged. A good objective has clearly defined limits and can robustly demonstrate whether it has sufficient resources allocated. A good objective will give sufficient confidence for the conclusion to rest only in the facts: there is sufficient evidence to say yes, or there is insufficient evidence to say yes, but that does not mean it is sufficient to say no; the risk can be ameliorated, but not removed.

A good objective, indeed any objective, is not an answer but a *question.*

The answer the objective may achieve might be definitive, but in our imperfect discipline it will more often be a position within a threshold of probability.[2]

---

1 Note the use of word 'appraisal' at this stage. This is because not all projects or operations require surveillance data, as will become clearer as the book progresses.

2 While yes is often achieved, no is rarely an option, for those of us working out of necessity will have limits imposed upon us from outside, and those limits will have a bearing on how ambitious our objective will be.

The first question that will logically be asked is: is surveillance necessary?

In order to answer that question, the objective threshold adopted in this book is that of *"reasonable likelihood"*.

"Reasonable likelihood" is enshrined in ODPM Circular 06/2005: Biodiversity and Geological Conservation – Statutory Obligations and their impact within the Planning System and within National Planning Practice Guidance.

ODPM Circular 06/2005

ODPM Circular 06/2005 states:

> *"The presence of a protected species is a material consideration when a planning authority is considering a development proposal that, if carried out, would be likely to result in harm to the species or its habitat."*

Therefore:

> *"It is essential that the presence or otherwise of protected species, and the extent that they may be affected by the proposed development, is established before the planning permission is granted, otherwise all relevant material considerations may not have been addressed in making the decision."*

However:

> *"Bearing in mind the delay and cost that may be involved, developers should not be required to undertake surveys for protected species unless there is "reasonable likelihood" of the species being present and affected by the development."*

National Planning Practice Guidance: Natural Environment – Biodiversity and Ecosystems

Paragraph 016 of National Planning Practice Guidance (NPPG): Natural Environment – Biodiversity and Ecosystems states:

> *"An ecological survey will be necessary in advance of a planning application if the type and location of development are such that the impact on biodiversity may be significant and existing information is lacking or inadequate."*

Furthermore:

> *"Where an Environmental Impact Assessment is not needed it might still be appropriate to undertake an ecological survey, for example, where protected species may be present."*

However:

> *"Local planning authorities should only require ecological surveys where clearly justified, for example if they consider there is "reasonable likelihood" of a protected species being present and affected by the development. Assessments should be proportionate to the nature and scale of the development proposed and the likely impact on biodiversity."*

"Reasonable likelihood": 'more-likely-than-not'

In the context of both pieces of guidance, *"reasonable likelihood"* means 'more-likely-than-not', which in fully accurate terms means >50% probability.

It does not, therefore, mean a vague potential, but rather a situation that is on the positive side of the 'probable' threshold, and probable requires proof; a foundation of robust scientific evidence.

Therefore, in an appraisal in support of a planning application, for every question we seek to answer we should provide proof on an increasing scale. Adopting the same principle in an amateur search would also be sensible, as it directs energy to where it is most likely to be rewarded. A list of example questions might include:

1. Is it 'more-likely-than-not' that tree-roosting bat species are in range of the site?
   Can we prove it?

2. Is it 'more-likely-than-not' the site holds trees?
   Can we prove it?

3. Is it 'more-likely-than-not' those trees hold Potential Roost Features (PRFs)?
   Can we prove it?

4. Is it 'more-likely-than-not' those PRF are suitable to hold roosting bats of the species that are 'more-likely-than-not' to visit the site?
   Can we prove it?

5. Is the combination of bat species, habitat and PRF such that it is 'more-likely-than-not' that bat roosts will be present?
   Can we prove it?

6. Is it 'more-likely-than-not' that the operation proposed will affect any tree-roosts that might be present within the Zone of Influence[3]?
   Can we prove it?

7. Is it 'more-likely-than-not' that existing information is lacking or inadequate to inform an Impact Assessment?
   Can we prove it?

8. Can we design surveillance that will be 'more-likely-than-not' to provide any missing or inadequate information that will be sufficient to inform an Impact Assessment? Fundamentally:

   a. Is it 'more-likely-than-not' the surveillance will encounter each individual species of bat or evidence of bats if they are present?

   b. Is it 'more-likely-than-not' that we will be able to conclusively identify any bats encountered?

   c. Is it 'more-likely-than-not' that we will be able to count any bats encountered?

   d. Is it 'more-likely-than-not' that we will be able to sex any bats encountered?

   And if not …

   e. Is it 'more-likely-than-not' that if the surveillance does *not* encounter bats, we will be able to interpret the field-signs we have observed to either perform a robust appraisal based on the data available or to support a request for more resources?

   And can we prove it?

   Can we demonstrate that our findings agree with a body of evidence to which the readers of our report have access and can see for themselves that, while the surveillance may not demonstrate presence conclusively, it does support a statement of probability that is above mere potential.

---

3 The geographical scale at which impacts upon a target species caused by an operation are identifiable.

Furthermore, can we demonstrate that the methods we have used and the effort we have expended in pursuing the objective were sufficient to support a robust conclusion?

And so, we come to proportionality.

1.5 PROPORTIONALITY

The need for a proportional approach is enshrined within National Planning Practice Guidance, which requires that:

> Assessments should be proportionate to the nature and scale of the development proposed and the likely impact on biodiversity.

Proportionality decides how much effort we expend pursuing the objective, and again we have the word 'likely'; not a vague potential, but a predictable probability that should be supported by tangible evidence.

The concept of proportionality might simply be thought of as a balance; a set of scales.

An example of proportionality might be the balance between the predicted magnitude of an impact in terms of timing, duration, extent and reversibility, versus the outcome of that impact (i.e. its effect). However, it might also be the balance between the loss of one area of habitat versus the gain of another, or the outlay in terms of appraisal effort versus the additional confidence the expense will provide.

In the latter scenario, in the context of an Impact Assessment, it might reasonably be argued that different bat species will warrant different levels of confidence and therefore different levels of expense.

To illustrate, different bat species have different population sizes and therefore make up different proportions of our bat fauna. In addition, different species have different population trends and different geographical ranges. The situation in respect of each species (including species that are not known to roost in trees) is summarised in Table 1.1.

Population estimates are taken from Bat Conservation Trust (2010) except Bechstein's bat *Myotis bechsteinii* which is taken from Harris *et al.* (1995). Status and population trend are taken from Bat Conservation Trust (2014). Broad distributions are taken from Harris and Yalden (2008), except Nathusius' pipistrelle *Pipistrellus nathusii* which is taken from www.nathusius.org.uk/Distribution.htm.

Even if the only difference between the two species were their population size, a proportionate approach in respect of Daubenton's bats *Myotis daubentonii* might arguably be significantly less in terms of outlay than the same in respect of barbastelles *Barbastella barbastellus*, simply because the risk of a negative effect upon the *population* of Daubenton's bats is lower.

For example, suppose a single area of woodland was to be felled in order for a development to proceed, and the woodland supported a single colony of Daubenton's bats and a single colony of barbastelles. For the sake of this discussion, both colonies number 100 bats. If we accept the figures set out in Table 1.1 as broadly accurate, the loss of that woodland would result in the displacement of 0.017% of the national population of Daubenton's bats, but potentially 2% of the barbastelle population.

Therefore, referring to ODPM Circular 06/2005, any operation that impacts upon habitat which even an individual colony of barbastelle is reliant will be more *'likely to result in harm to the species'* than the same operation would upon a colony of Daubenton's bats, and there is therefore a *'reasonable likelihood of the* species *being ... affected by the development'*.

Table 1.1 The conservation status of British bat species, ordered in increasing level of significance

Species	UK population estimate and proportion of bat fauna	UK status	Broad distribution
Common pipistrelle	2,430,000/49.41%	Common and increasing	All regions
Soprano pipistrelle	1,300,000/26.43%	Common and stable	All regions
Daubenton's bat	560,000/11.39%	Common and increasing	All regions
Brown long-eared bat	245,000/4.98%	Common and stable	All regions
Natterer's bat	148,000/3.01%	Common and increasing	All regions
Whiskered bat	64,000/1.3%	Uncommon but stable	All regions but Scotland
Noctule	50,000/1.02%	Uncommon but stable	England
Brandt's bat	30,000/0.61%	Uncommon but stable	Republic of Ireland and England
Leisler's bat	28,000/0.57%	Uncommon and trend unknown	Ireland and patchy in central and southern England
Lesser horseshoe-bat	18,000/0.37%	Rare but increasing	Wales and south-west England
Nathusius' pipistrelle	16,000/0.33%	Uncommon and trend unknown	All regions
Serotine	15,000/0.3%	Uncommon but stable	Wales and south-west, southern and eastern England
Greater	>6,600/0.13%	Very rare but increasing	Wales and south-west England
Barbastelle	5,000/0.1%	Rare and trend unknown	Wales and England
Bechstein's bat	1,500/0.03%	Very rare and trend unknown	South-west and southern England
Grey long-eared bat	1,000/0.02%	Very rare and trend unknown	Southern England
Alcathoe's bat	No data	Unknown	Southern England

However, although we have already identified a disparity in risk, this example is oversimplified and assumes both species have the same population trend, the same range, require the same quantity and quality of habitat, are equally detectable, and are equally identifiable, but they are not; the rarer species confounds on all fronts. Furthermore, whilst there is evidence to support the statement that Daubenton's bats readily adopt artificial roost boxes in the active period, and will overwinter elsewhere in a subterranean site, there is a paucity of evidence (all of which is anecdotal) in respect of compensatory action relating to the barbastelle, which may well remain within the woodland to overwinter. The barbastelle therefore warrants greater effort from the outset, and the proportionality balance for the two species will logically be weighted differently (as it would for all species).

This approach to proportionality was legally tested in the context of planning in 2014, in the case of *Cheshire East Council v Rowland Homes* [2014] EWHC 3536 (Admin). In this case, a preliminary ecological survey had been undertaken of a site proposed for development on which there were previous records of bats. However, no further specific bat surveys were undertaken. Yet the court held (for a development of just under 100 homes) that the inspector had nevertheless discharged, lawfully, his regulation 9(3) duty when he granted full planning permission for the development, whilst concluding (in the absence of survey data) that there was adequate space within the site to accommodate such protected species and, as derogation would not be detrimental to the maintenance of the populations of the species concerned at a favourable conservation status in their natural range, there were no grounds to suggest derogation under licence would be refused if bats were found to be present at the safeguarding stage.

This site was in Cheshire and there were therefore no grounds to suggest a *"reasonable likelihood"* that a rare species would be present. However, the site had held a farmhouse, a number of brick-built agricultural buildings, a garden and associated farmyards, so the potential for roosting bats to occur could not be ignored.

More recently, Natural England has published the results of a public consultation in respect of European Protected Species Licensing (EPSL) Policy (Natural England 2016). The outcome of the consultation in respect of Policy 4 – *Appropriate and relevant surveys where the impacts of development can be confidently predicted* was that Natural England may accept a lower than standard survey effort in support of a planning application where *all three* of the following apply:

1. *The costs or delays associated with carrying out standard survey requirements would be disproportionate to the additional certainty that it would bring;*

   and

2. *The ecological impacts of the development can be predicted with sufficient certainty;*

   and

3. *Mitigation or compensation will ensure that the licensed activity does not detrimentally affect the conservation status of the local population of any European Protected Species.*

The case law and Natural England's EPSL Policy 4 therefore appear to indicate that where it can be proven that effective mitigation and compensatory measures exist for each bat species for which there is a *"reasonable likelihood"* of presence, but not of an effect upon the *species* conservation trend, it may not be proportionate for the developer to have to perform detailed surveillance in advance of a planning application. In such a situation, a reasoned and evidence-supported desktop appraisal, with an accompanying due-diligence safeguarding strategy,[4] might therefore be sufficient.

The decision whether, and exactly when surveillance is required is not therefore straightforward, but the four stages of a legislative proportionality test set out by Craig and de Búrca (2015) offer a useful framework in deciding what level of appraisal would be reasonable in the context of different magnitudes of impact and their predicted effects

---

4   In practice, it would be likely that the need to assess the status of roosting bats within the site prior to the destruction or disturbance of any potential roost habitat would be compelled under a Planning Condition (typically within an overarching Ecological Management Plan).

upon individual bat species. Adapting the test to suit the context of this book, the four stages that are considered are as follows:

1. There must be a specific and identifiable aim for the appraisal proposed.
2. The appraisal method must be suitable to achieve the aim (potentially with a requirement of evidence to show it will have that effect).
3. The method and level of effort must be necessary to achieve the aim, that there cannot be any cheaper way of doing it.
4. The measures must be reasonable, i.e. suit the circumstances, rarity of the species, conservation trend, etc.

This is where this chapter ends. From hereon, **Step 1** will be our objective. In this book we have adopted *"reasonable likelihood"*; is it 'more-likely-than-not' that tree-roosts might be present?

It is beyond the scope of this book to attempt to anticipate all the variables in terms of timing, duration, magnitude and reversibility of impacts that each individual appraisal might encompass. However, we can review the ecological evidence, and summarise it into a framework that may be practically applied in the performance of a dynamic risk-assessment, as to whether there are grounds to conclude it is 'more-likely-than-not' individual bat species might roost within our site and might thereby be present to be affected. This is actually not as difficult as it might first appear, although summarising the evidence and setting out the framework will take five chapters to achieve (Chapters 2–6).

Step 2 will be dealt with in Chapter 7, where we identify and summarise the methods available to us, as well as their strengths and weaknesses.

Step 3 will be dealt with in Chapters 8–11, where we match the behaviour of each species with the suite of methods we can employ in our site, to identify the single most effective method for each of the individual species, and look closely at the surveillance effort required to reward with robust data in each season.

Step 4 feeds back into Chapter 1, where we initially considered the rarity of each bat species and their conservation trends. However, whether the outlay is reasonable will require consideration on the part of the individual surveillance team that is beyond the scope of this book.

Finally, in Chapter 12 there is some experience-based advice regarding managing client expectations in crisis situations, hazardous trees, bad timing and where surveillance teams might get a second opinion if they are uncertain their strategy is robust.

The evidence base from which this text will primarily draw comprises three sources:

» Andrews H *et al.* 2016. *Bat Tree Habitat Key*, 3rd edn. Bridgwater: AEcol.
» Andrews H and Gardener M 2016. *Bat Tree Habitat Key – Database Report 2016*. Bridgwater: AEcol.
» The Bat Tree Habitat Key Database.

All are available free of charge from www.battreehabitatkey.com.

Save where the information is individually cited and appropriately referenced, it may be assumed that the evidence base is one or more of the three primary sources listed above.

Despite this evidence-based foundation, every situation encountered is a mobile point within a range on a sliding scale; there are no absolutes. In the context of advice, however, boundaries must nevertheless be defined and the threshold of *"reasonable likelihood"* (i.e.

'more-likely-than-not', and in mathematical probability terms >50%) is the threshold that has been adopted in this book.

Therefore, the categorisations identified will represent the *typical* situations in which each bat species has been encountered. It must also be accepted that the categorisations rely not only upon the data gathered and made available to date, but also upon conspicuous holes in the data. As a result, some of the categories rely on exceptions that prove rules.[5]

In order to use this threshold in plain English, *"reasonable likelihood"* is presented as 'more-likely-than-not' in the text generally. However, where sections culminate in data interpretation that will decide whether or not further outlay would be proportionate, the conclusions are presented as a 'test of *"reasonable likelihood"'*. For complete accuracy, in the tabulated results of statistical analysis, the threshold is any situation where the result is >50%.

---

5   An example of this might be that despite the massive amount of radiotracking performed both in the UK and on the continent, there remains a paucity of evidence relating to bats roosting in ivy *Hedera helix*, which supports the suggestion that the presence of a roost within ivy is below the threshold of 'more-likely-than-not'.

CHAPTER 2

Tree-Roosting Bats

In this chapter	
Tree-roosting bat species	A review of the bat species that have been recorded roosting in trees
Wooded habitat and tree species	The broad environments created and occupied by trees and which bat species occupy them
Sensitivity to isolation	How far individual species may cross open ground to reach isolated areas of wooded habitat and individual trees
Seasonal tree-roost occupancy and roost size	When individual bat species occupy trees and how many bats might be present
Roost heights	The heights of tree-roosts occupied by bats in different seasons
General tree-roost preferences	Which bat species favour voids, crevices or both

2.1 Tree-roosting bat species

In the British Isles, 14 bat species have been recorded roosting in trees. Grouping the tree-roosting species into their six families, we have:

» The barbastelle *Barbastella barbastellus*.
» The *Myotis* sp., which comprise:
  – Bechstein's bat *Myotis bechsteinii*;
  – Alcathoe's bat *M. alcathoe*;
  – Brandt's bat *M. brandtii*;
  – Daubenton's bat *M. daubentonii*;
  – Whiskered bat *M. mystacinus*; and
  – Natterer's bat *M. nattereri*.
» The *Nyctalus* sp., which comprise:
  – The noctule *Nyctalus noctula*; and
  – Leisler's bat *N. Leisleri*.
» The *Pipistrellus* sp., which comprise:
  – Nathusius' pipistrelle *Pipistrellus nathusii*;
  – The common pipistrelle *P. pipistrellus*; and

– The soprano pipistrelle *P. pygmaeus*.

» The brown long-eared bat *Plecotus auritus*; and

» The lesser horseshoe-bat *Rhinolophus hipposideros*.

2.2 Wooded habitat and tree species

Although all the 14 bat species identified occupy trees, different combinations of those bat species occupy different habitats, and these habitats can be divided into six broad wooded-habitat types, comprising:

» Woodland.

» Plantation.

» Traditional orchard.

» Wooded linear landscape elements, encompassing:

– Hedges;

– Shelter-belts; and

– Riparian fringes.

» Parkland, encompassing:

– Wood-pasture; and

– Pleasure-grounds.

» Churchyards and gardens.

Applying the threshold of 'more-likely-than-not' excludes consideration of trees isolated within tillage, because the BTHK Database holds no record of a tree-roost in this context, nor were any such roosts identified by Andrews *et al.* (2016). In addition, the BTHK Database holds only one record of a tree-roost that is isolated within pasture, but this record provides two valuable lessons in habitat assessment. The individual record comprises an impressive lightning-strike on a pedunculate oak *Quercus robur* occupied by a barbastelle maternity colony. The tree is 60 m from a recently grubbed-out hedgeline and within 100 m of a wooded pond in what was until very recently a long-established wood-pasture associated with both the Grand Western Canal and River Tone. The roost feature, historic context and remaining habitat component may all reasonably be predicted to have a bearing on the appeal of the situation.

The first lesson is that an investigation into the longevity of the wooded landscape, using historic maps and aerial photographs, may yield useful insights as to what the habitat really is: isolated tree, or degenerating parkland in association with small fields divided by hedges.

The second lesson is that some bat species have a restricted range of favoured Potential Roost Features (PRFs), and the barbastelle is one of them; lightning-strikes are prized above all other PRFs by maternity groups, with bark a close second (although lifting bark of the nature exploited by barbastelles may be the result of a lightning-strike that killed the tree). If an absolutely eye-popping PRF exists on a significantly old tree in a 'sub-optimal'[1] situation, a colony may have adopted it when the situation was 'optimal'

---

1   Sub-optimal simply means less than perfect. There are few perfect situations in nature, and even where they exist they have good and bad periods so the range may be moderately wide. Just because something was sub-optimal this year does not mean it always was, or always will be.

and have refused to leave it, or have been unable to simply because no other suitable PRF has come about that might offer an alternative.

Different tree species are more or less abundant in different situations and form different canopy communities. The different characteristics of the individual tree species, combined with the different environmental conditions in which they occur, result in them forming different PRFs in different abundances. For example, wood decay fungi are typically more abundant in ash *Fraxinus excelsior* than oak woodland, but a lowland pedunculate oak woodland will typically have more infected trees than an upland sessile oak *Q. petraea* woodland. Turning that on its head, an upland sessile oak woodland will typically hold more structurally damaged trees than a lowland pedunculate oak woodland, but a lowland oak woodland will still hold more damage PRF than a lowland ash woodland.

However, the final point that should be borne in mind from the outset is that not all tree species are gregarious and create woodland. Some of the species that readily form PRFs are not woodland trees, such as the white willow *Salix alba* and crack willow *Salix fragilis*. In addition, some of the woodland species that had historically been dominant across the British Isles following the last ice age are no longer dominant in the woodland canopy, such as the small-leaved lime *Tilia cordata*, large-leaved lime *Tilia platyphyllos* and aspen *Populus tremula*. Lastly, one tree species, the sycamore *Acer pseudoplatanus*, which forms PRFs when young, is often selectively eradicated from woodlands as it is considered a weed. This last species is in fact of such high value to tree-roosting bats that it is considered separately at the close of this section.

2.2.1 Woodland

Woodland can be divided into those that have a specific purpose (i.e. the growing of timber and underwood) and secondary woodland (i.e. woodland that is due to succession on neglected farmland, moorland, etc.). In a woodland the canopy is for the most part continuous but may include glades, and often a secondary, lower shrub-layer canopy or thicket.

Rodwell (1991) identifies 19 woodland communities in which the dominant species encompass nine families, comprising:

1. Willows *Salix* spp., comprising:
   a. W1 – Lowland grey willow *Salix cinerea*;
   b. W2 – Lowland grey willow, downy birch *Betula pubescens* and alder *Alnus glutinosa*; and
   c. W3 – Northern lowland bay willow *Salix pentandra* and grey willow.
2. Birch *Betula* sp., comprising:
   a. W4 – Lowland and upland fringe downy birch.
3. Alder, comprising:
   a. W5 – Lowland alder with grey willow and ash;
   b. W6 – Lowland alder; and
   c. W7 – Upland alder with ash.
4. Ash, comprising:
   a. W8 – Southern lowland ash and field maple *Acer campestre*; and
   b. W9 – Northern lowland ash, rowan *Sorbus aucuparia* and downy birch.
5. Oaks *Quercus* spp., comprising:
   a. W10 – Lowland pedunculate oak and silver birch *Betula pendula*;

b. W11 and 16 – Upland fringe sessile oak and downy birch; and

c. W16 – Southern lowland oak (both *Quercus robur* and *petraea*) and birch (both *B. pubescens* and *pendula*).

6. Beech *Fagus sylvatica*, comprising:

a. W12 – Beech with horse chestnut *Aesculus hippocastanum* and silver birch; and

b. W14 and 15 – Southern lowland beech with pedunculate oak.

7. Yew *Taxus baccata*, comprising:

a. W13 – Yew with ash.

8. Scots pine *Pinus sylvestris*, comprising:

a. W18 – Highland native pine woodland.

9. Juniper *Juniperus communis* ssp. *communis*, comprising:

a. W19 – Juniper with downy birch and rowan.

2.2.2 Plantation
Plantations are characterised by deliberate planting of a 'timber' crop in parallel ranks. Typified by trees of all one age group, either a single species (or if others occur, one is by far the dominant species), and one height, in a closed-canopy with little or no shrub-layer.

In the plantation context, the trees can be separated into their two Divisions:

» **Conifers**, comprising: Sitka spruce *Picea sitchensis*; Norway spruce *P. abies*; European larch *Larix decidua*; Japanese larch *L. kaempferi*; Douglas fir *Pseudotsuga menziesii*; western hemlock *Tsuga heterophylla*; western red cedar *Thuja plicata*; Scots pine; and Corsican pine *Pinus nigra*.

» **Broadleaves**, comprising: sycamore; ash; pedunculate oak; sessile oak; sweet chestnut *Castanea sativa*; beech; and hybrid poplars *Populus* spp.

2.2.3 Traditional orchard
Traditional orchards are characterised by typically small enclosures containing a mono-culture of fruit-trees, planted in straight lines.

Natural England (2010a) describes traditional orchards as including subspecies of domestic apple *Malus domestica*, pear *Pyrus communis*, cherry *Prunus avium*, and plum *Prunus domestica* and typically planted on free-draining soil in situations where they would not be exposed to extremes of rainfall, wind or cold. The stand is typically all of one age, and may comprise trees that are all over 100 years old (Natural England 2010b), set in equidistant to the boundary vegetation and each other, with spacing typically at *c.* 8 m for apple, pear and cherry within a range of *c.* 3–20 m (JNCC 2008). The result of the spacing is a loose, low canopy through which sufficient light can penetrate to support a grassland sward upon which sheep may be grazed. The even age and grazing results in the canopy being parallel both on top and underneath with a clearly defined browse-line.

2.2.4 Wooded linear landscape elements
Wooded linear landscape elements encompass three distinct situations with gradations in canopy continuity, comprising:

» Hedges – characterised by trees of similar ages at wide spacing and often without canopy continuity.

» Shelter-belts – characterised by trees of a similar age at close spacings, with a continuous, thin and relatively high canopy, but relatively short overall section length.

> » Riparian-fringe – characterised by trees of all age ranges at close spacings with a continuous low and wide canopy, typically in short sections, with relatively short gaps in between each section (although sections may be on opposite banks).

The factor that connects these habitats is just that: connection. Although much of a hedge may be unsuitable for roosting, the feature provides a navigable landmark, as do lines of trees. Even where trees are widely spaced on river-, stream-, canal- and even lake-banks, the watercourse represents a reliable route, and emergent vegetation may be sufficient to provide cover equivalent to that provided by a hedgerow.

Hedges

Hedges are typically low and narrow: one or two shrubs deep, giving managed dimensions of 1.5–2 m height and width and outgrown dimensions of 5–8 m height and 3–5 m width. Hedgerows may divide farmland, or bound footpaths, bridleways and highways. Regardless of the situation, where they occurred, standard trees were set in at sufficient spacing (typically, 10–30 m apart) to allow a wide canopy to be attained at maturity. Where this was managed correctly, the result would have been a linear canopy over a dense linear shrub-layer beneath. However, whilst very little harvesting appears to have occurred, even where a crop of timber trees had initially been established, due to later neglect and mismanagement it is now more common to find significant gaps between the canopies of the trees that remain.

Pollard *et al.* (1974) summarise data from a census of hedgerow timber performed by the Forestry Commission in 1951 which found the most common hedgerow trees to be oaks, followed (in order of frequency) by the common elm *Ulmus procera*, ash, beech and sycamore.

Shelter-belts

Shelter-belts may be planted for reasons such as:

> » To provide a wind-break for tillage or horses.
> » For the dual purpose of providing game-cover and an obstacle over which to drive the birds above the waiting guns.
> » As visual screening.

In all situations, the mature shelter-belt will comprise full-grown trees and will typically exist as a component within a hedgerow network. Regardless of the situation, shelter-belts are not usually managed for their timber.

Where the objective is to provide a wind-break, the belt may comprise a single rank of one species (e.g. Lombardy poplar *Populus nigra* var. *italica* or Leyland cypress *Cupressus × leylandii*). Where it is to provide game cover, it is common to find two to three staggered ranks comprising a mix of native tree species but over a conspicuously 'ornamental' shrub-layer (i.e. snowberry *Symphoricarpos albus*, cherry laurel *Prunus laurocerasus* and rhododendron *Rhododendron ponticum*). In screening situations, the species mix may well be eclectic and comprise a mix of both native and ornamental trees and shrubs.

Riparian-fringe

The riparian-fringe category comprises mature bankside standard trees including crack willow, white willow, alder and, less frequently, ash and even oak, in some instances over an untidy shrub-layer of smaller willow species.

2.2.5 Parkland

Parkland is typified by a situation where the growing of trees is subordinate to, or associated with, another land use. In the context of this book, parkland includes wood-pasture and landscaped pleasure-grounds of significant extent (including golf courses).

Regardless, the habitat in both situations is characterised by individual and small groups of mature and over-mature trees at wide spacings.

These two wooded habitats encompass a loosely wooded landscape that may be of significant age and in which the trees may have been present in specific situations for generations.

Wood-pasture

Wood-pasture can be broadly divided into two forms, comprising:

» **Lowland wood pasture:** Derived from medieval hunting forests (Harmer *et al.* 2010) and typified by large maidens over sheep-grazed pasture within sight of a stately home.

» **Upland wood pasture:** Which historically provided shelter for grazing animals and wood for rural communities (Harmer *et al.* 2010), typified by pollard oaks.

In wood pasture the trees are typically pollard oaks, cut periodically at 3–4 m height in order that the new growth escapes browsing by livestock (Rackham 1995).

Pleasure grounds

Pleasure grounds typically contain both broadleaved and coniferous species, the latter of which will in many cases be entirely unrecognisable from their plantation brethren; for example, Edlin (1976) noted that when grown in open parkland and allowed to reach maturity, pines develop a rugged structure with irregular upper branches, similar in outline to a broadleaved tree. Standard specimens of species that are frequently encountered comprise sweet chestnut, black locust *Robinia pseudoacacia*, oaks, common lime *Tilia × europaea*, yew (particularly where it forms clumps), cedars *Cedrus* spp. and the giant redwood *Sequoiadendron giganteum*.

2.2.6 Churchyards and gardens

Churchyards

As with parkland, churchyards may have changed very little for hundreds of years and standard trees within them are often of significant age. Species that are frequently encountered are the common lime and yew.

Rural gardens

Rural gardens are essentially divided parkland but subject to far more disturbance in the form of artificial lighting (both static domestic and street-lighting and mobile vehicle lights), human presence and, importantly (because they can climb), domestic cats. Nevertheless, unless it is obvious that a housing estate was built round the trees, a large garden that encompasses mature trees within landscaping may be of significant age. Species that are worthy of note in this context are the oaks, ash, common lime and Monterey cypress *Cupressus macrocarpa*. However, trees in even rural gardens may find themselves stressed by drought and landscaping, and damaged by inappropriate or incompetent surgery. As a result, any tree in such a context is worth close-inspection.

2.2.7 Sycamore

Sycamore warrants individual consideration.

Sycamore occurs as a mature tree in nine of the woodland communities identified by Rodwell (1991), i.e. W6, 8, 9, 12, 13, 14, 15, 16 and 17. Furthermore, although outside small areas of coppice it is never dominant, it is one of the most frequently encountered broadleaved weed species in plantations, and Pollard *et al.* (1974) also note its frequency as a hedgerow tree in Wales and Scotland.

Save where it is targeted in the genocidal fervour of purist management in a futile attempt at eradication, it is typically ignored; upstaged by the larger ash and oak (often themselves genetically modified and a good deal less 'native' or natural than this accused alien). Under stressed conditions, and where it is subject to bark-stripping by grey squirrels *Sciurus carolinensis*, the young sycamore is typically stunted, with a deformed and contorted stem, and limbs often presenting acute angles.

Due to the disease and damage resistance of a good deal of the oak and ash within 'semi-natural' woodlands, the crippled offspring of this weed species often holds a superabundance of PRFs. The BTHK Database holds records of sycamore roosts occupied by barbastelle, Daubenton's bats, Natterer's bats, common pipistrelle, soprano pipistrelle and brown long-eared bats. Furthermore, sycamore is occupied year-round for all purposes, and these trees have a diameter at breast height ranging from 8.8 cm upwards.

Referring back to the population estimates given in Chapter 1 suggests that these six bat species make up roughly 95% the UK's bat fauna, and we may therefore confidently suggest that it is 'more-likely-than-not' that tree-roosts in sycamore are not uncommon. In fact, a single compartment can contain a staggering number of occupied stems on a single day, at any time of year.

The empirical evidence that proves sycamore is one of the most important tree-roost species is already significant; ignore it at your peril.

2.2.8 Summary
Table 2.1 summarises the six broad wooded habitats in which roosts of the individual tree-roosting species have been recorded. Importantly, Table 2.1 also identifies the habitats in which there are insufficient grounds to suggest that a roost of an individual species would be 'more-likely-than-not'.

2.3 Sensitivity to isolation
The preference for commuting in the shelter of linear landscape elements, exhibited by all but the noctule and Nathusius' pipistrelle, might suggest that a parcel of wooded habitat that was isolated in the landscape would not be 'more-likely-than-not' to be exploited for roosting, even where the habitat was otherwise suitable and held PRFs.

However, there is a difference between reliance upon linear landscape elements, and a preference for commuting along them; regardless of the broad preferences, gaps are crossed by a significant number of species as demonstrated at Table 2.2.

Therefore, in the absence of another contributing factor (i.e. lighting which has been shown to have a detrimental effect upon movements – see Stone 2011, 2013; Stone *et al.* 2012), caution should be exercised before scoping out an otherwise superficially suitable habitat parcel due to perceived isolation.

2.4 Seasonal tree-roost occupancy and roost sizes
Different bat species are present in trees in different seasons and in different numbers.

2.4.1 Seasons
The annual tree-roost pattern of occupancy can be broadly divided into six periods as follows:

Winter – January and February.

Spring flux – March and April.

Pregnancy – May and June.

Nursery – July and August.

Mating – September and October.

Autumn flux – November and December.

Table 2.1 The wooded habitats within which individual bat species have been recorded

	Woodland	Plantation	Traditional orchard	Linear landscape element	Parkland	Churchyard and garden
Barbastelle	✓	—	—	—	✓	—
Bechstein's bat	✓	—	✓	Hedge: late summer/autumn	—	—
Alcathoe's bat	✓	—	—	—	—	—
Brandt's bat	✓	—	—	—	✓	—
Daubenton's bat	Near water	—	✓	Riparian-fringe	✓	—
Whiskered bat	✓	—	—	—	—	—
Natterer's bat	✓	✓	✓	Hedge/treeline: late summer/autumn	✓	—
Leisler's bat	✓	—	—	—	✓	—
Noctule	✓	—	—	✓	✓	✓
Nathusius' pipistrelle	Near still water	Near still water	—	Riparian-fringe	—	—
Common pipistrelle	✓	—	—	✓	✓	✓
Soprano pipistrelle	✓	—	—	—	✓	—
Brown long-eared	✓	✓	—	Hedge/treeline: late summer/autumn	—	—
Lesser horseshoe	✓	—	—	—	✓	—

These periods reflect the situation in respect of when individual species are present in trees but also incorporate a buffer to account for anomalous years. Thus, there are two flux periods, spring and autumn, which (within reasonable limits) consider annual fluctuations in the severity of winters and the onset of spring, and therefore the month that the species that overwinter in subterranean habitats are first and last recorded in trees.

The barbastelle, Natterer's bat, Leisler's bat, noctule, common pipistrelle, soprano pipistrelle and brown long-eared bat use trees year-round.

Bechstein's, Alcathoe's and Daubenton's bats use trees during the summer for giving birth and raising young, but although all these species may remain into the autumn, they have not been recorded overwintering in trees in the British Isles. The presence of all three species in trees is erratic in the spring and from the mating season into the autumn flux.

Table 2.2 Tolerance to habitat isolation/fragmentation exhibited by tree-roosting bats

Species	Tolerance to isolation/fragmentation
Barbastelle	Moves freely across open landscapes after dark, crossing major roads and wide estuaries (Zeale *et al.* 2012). Continental migrations range from 10 to 290 km (Hutterer *et al.* 2005)
Bechstein's bat	Has been recorded flying directly across fields to reach foraging habitat and to return to roosts at dawn (Harris and Yalden 2008). Non-migratory (Hutterer *et al.* 2005)
Brandt's bat	Crossed 180 m clear-fell upon leaving roost (de Jong 1995), and although they generally avoid open habitats, tracked bats crossed open ground when returning to a roost before dawn (Schaub and Schnitzler 2007). Continental migrations range from 10 to 618 km (Hutterer *et al.* 2005)
Daubenton's bat	Followed linear landscape elements when leaving the roost at dusk, but not when returning at dawn (Schaub and Schnitzler 2007). Crossed a gap *c.* 200 m wide between two small groups of mature trees (Downs and Racey 2006). Maximum migration 19 km in UK (Hutterer *et al.* 2005)
Whiskered bat	Uses open landscapes and forages over large waterbodies (Dietz *et al.* 2011), which suggests the species is not confined to closed habitat. Continental migrations of 10–70 km not uncommon (Hutterer *et al.* 2005)
Natterer's bat	Forages over freshly mown meadows (Dietz *et al.* 2011), which suggests the species is not confined to closed habitats. Considered a facultative migrant (Hutterer *et al.* 2005)
Leisler's bat	Crosses and forages within open areas (Dietz et al. 2011). Maximum migratory distance: 1,567 km (Hutterer *et al.* 2005)
Noctule	Almost all landscape types used and forages in open areas (Dietz *et al.* 2011). Maximum migratory distance: 1,546 km (Hutterer *et al.* 2005)
Nathusius' pipistrelle	Often forages along linear structures, e.g. firebreaks, forest edges but is migratory (Hutterer *et al.* 2005) and therefore crosses wide distances of open habitat (i.e. the North Sea)
Common pipistrelle	Recorded equally in areas with a network of commuting routes and those without (i.e. open ground), no discernible effect on presence brought about by the fragmentation, concluded that 'open areas of 110–150 m wide do not form a serious barrier' (Verboom and Huitema 1997). Crossed a gap approximately 200 m wide between two small groups of mature trees (Downs and Racey 2006). Followed linear landscape elements when leaving the roost at dusk, but not when returning at dawn (Schaub and Schnitzler 2007). Regular movements of 34 km recorded (Hutterer *et al.* 2005)
Soprano pipistrelle	Forages widely over lakes and rivers and will cross large open areas (Downs and Racey 2006)
Brown long-eared bat	Crossed dual carriageways *c.* 30 m wide, despite high traffic flow (Berthinussen and Altringham 2012)
Lesser horseshoe-bat	Crossed distances of 30–100 m between individual trees, and one bat used a flight-path over 200 m of open habitat through long-established parkland (Downs *et al.* 2016)

Colonies of Brandt's and whiskered bats do very occasionally exploit trees in the late spring and into the pregnancy season, but usage is by no means common and the BTHK project has not had sight of any evidence to support any suggestion that either species gives birth or suckles young in trees. Furthermore, although usage of trees during the mating and autumn-flux periods is not unheard of on the continent, the BTHK project has not had sight of any evidence that the same is true in the British Isles.

Although the common and soprano pipistrelle use trees year-round, only very exceptionally do females occupy trees when giving birth or when suckling young.

As for the Nathusius' pipistrelle, on the Continent, the species certainly does mate in trees in the spring and autumn and has been recorded overwintering in trees in the British Isles, but there are no records of a maternity colony or even an individual male occupying trees in the pregnancy or nursery seasons, and whether they mate in trees in the British Isles is uncertain.

Finally, thus far all the records of lesser horseshoe-bats occupying trees relate to pairs of bats that were encountered in the period between June and December.

2.4.2 Roost sizes

Males and females of different species behave differently in different seasons.

Winter

In winter, all but three species roost individually in their own roost-feature, the exceptions are the *Nyctalus* sp. which may stay in small groups, and the soprano pipistrelle which may overwinter in groups with over a hundred bats in one feature in a single tree.

Flux periods

In both the flux periods the numbers of bats in an individual roost can vary wildly but as yet there is insufficient data to support estimation of what they typically are.

Pregnancy and nursery

Up until relatively recently, it was still assumed that in the pregnancy and nursery periods the typical situation was for females to group together, while the males would continue to roost singly. However, while roosting alone appears to be more frequent than grouping in the winter, spring-flux and autumn-flux periods, and the BTHK Database holds records of individual barbastelle, Bechstein's' bat, Daubenton's bat, Natterer's bat, noctule, common pipistrelle, soprano pipistrelle and brown long-eared bats in the pregnancy and nursery periods, three of these species, plus Leisler's bat, have also proven to form all male groups, as follows:

» **Daubenton's bat** – males have been reported aggregating in bat-boxes by August *et al.* (2014) and on the Continent, where Encarnação *et al.* (2005) recorded all-male groups numbering up to *c.* 51 bats.

» **Natterer's bat** – all-male groups numbering up to 25 bats have been recorded on the Continent (Dietz *et al.* 2011).

» **Leisler's bat** – all-male groups of up to 12 bats have been recorded on the Continent (Dietz *et al.* 2011).

» **Noctule** – all-male groups of up to 20 bats have been recorded in the UK (Whitaker 1905) and on the Continent (Dietz and Keifer 2016).

This behaviour is worth bearing in mind for two reasons:

1. Colonies may occupy far fewer trees – if we think Bechstein's females are in two–three trees at any one time, and the males are in individual trees, then we might have up to 53 trees occupied on any one day (50 females in small groups divided between three trees, and individual males occupying 50 individual trees), but with our Daubenton's, Natterer's, Leisler's and noctules we may have only two trees occupied – all the males in one, and all the females in another.

2. It cannot be assumed that a group of bats encountered in May–August is a maternity group because they could all be males.

2.4.3 Mating

A paucity of data means that at present our estimations of the numbers of individual species groups in trees are still vague. In truth, no estimation is yet possible for the barbastelle; all that can be said with any certainty is that individuals have been recorded in trees within this period. The mating ecology of Alcathoe's bat is similarly a mystery. Daubenton's bats have been recorded in pairs during the mating season, whereas Natterer's bats are recorded in moderately large groups throughout September, before suddenly separating with only individuals recorded in trees from October on. Male Leisler's, noctules and Nathusius' pipistrelles take up residence in trees and sing from a static location, and male common and soprano pipistrelles perform song-flights around their roost-trees to attract passing females and form small to moderate-sized harems. Male Nathusius' pipistrelles have been shown to aggregate in two or three roosts in close groups in particularly favourable situations, but with wide gaps between these groups (Jahelková and Horáček 2011). Male brown long-eared bats roost individually and the females continue to group together in trees during the day, with the two sexes meeting in swarming sites at night (Furmankeiwicz 2016).

2.4.4 Summary

Table 2.3 is an evidence-supported summary of seasonal occupation, with the months that each species uses trees identified with green shading, and the numbers of bats typically present.

Table 2.3 The bat year

	Winter		Spring-flux		Pregnancy		Nursery		Mating		Autumn-flux	
	Jan	Feb	Mar	Apr	May	Jun	Jul	Aug	Sep	Oct	Nov	Dec
Barbastelle	1	1	1–?		F: c. 10–55/M: 1				1–?	1	1	
Bechstein's				1–?	F: c. 20–60/M:1				1–6		1–?	
Alcathoe's				?	F: c. 30–40/M: ?				?		?	
Brandt's				1	c. 35–95							
Daubenton's			1	1–?	F: c. 15–30/M: 1–50				1–2	1	1	
Natterer's	1	1		1–?	F: c. 15–25/M: 1–25				1–25	1	1	
Whiskered				1	c. 1–20							
Leisler's	1–?	1–?	1–?	1–?	F: c. 20–50/M: 1–12				M: 1/F: 1–10		1–?	
Noctule	1–?	1–?	1–?	1–?	F: c. 10–95/M: 1–20				M: 1/F: 1–18		1–?	
Nathusius'	1–2	1–2	1–2	1–?					M: 1/F: ?		1–2	
Common	1	1		1–?			1		M: 1/F: 1–3		1	
Soprano	1–100+	1–100	1–100	1–?			1		M: 1/F: 1–3		1–100+	
B. l-eared	1	1		1–8	F: c. 10–30/M: 1				1–20	1–2	1	
L. horseshoe							1–2		1–2		1–2	

Darker green illustrates periods in which it is 'more-likely-than-not' the species will be present roosting in trees. The lighter green illustrates periods in which the species has been recorded roosting in trees, but encounters are less common and presence is considered less reliable. Figures given represent the numbers of bats that might be expected. Where accounts are given for the individual sexes these are shown as F: female; and M: male respectively. Where the number ranges from 1–? this means that individuals and small groups have been encountered but the exact numbers could not be established (typically because the bats were torpid and did not emerge)

It is worth identifying at this point that some roost features are occupied in all seasons but by different bat species, like a time-share. Others are occupied only in one period for a short duration each year, by one colony/individual. Nevertheless, records on the BTHK Database provide evidence that the resident bat colonies of many species follow an ordered circuit of favoured trees each year, returning and leaving at roughly the same time.

2.5 Roost heights

Roosts at different heights are occupied in different seasons.

The structure of wooded habitat was broadly stratified into four zones by Simms (1971), as follows:

1. **Ground-layer:** 0–0.5 m.
2. **Field-layer:** 0.51–2 m.
3. **Shrub-layer:** 2.01–5 m.
4. **Canopy-layer:** 5.01–30 m.

Although these zones are more normally applied to woodland and plantation situations, application in the field has demonstrated that even when growing in an open situation, the crown of an individual tree may be sufficiently dense even down to ground level that the canopy embodies field, shrub and canopy cover.

At present, the records held on the BTHK Database only provide meaningful insight into the heights of roosts occupied when all bat species are combined, and this only allows resolution within three scales, these being: one bat; two–four bats; and five or more bats. Even accepting this 'lumping', it should be borne in mind that there will be considerable bias to the height data for two reasons: (i) significantly more sampling takes place in the ground and field-layers than currently does at height; and (ii) there is a bias for sampling in the warmer and drier months across all heights.

However, despite this currently low resolution and accepting the significant bias in height and seasonal sampling, we can state with certainty that roosting bats are present in the field, shrub and canopy-layers year-round, and these roosts may be occupied in the winter and both the spring- and autumn-flux periods.

2.5.1 Winter

The BTHK Database holds records in all the layers during winter, with an individual soprano pipistrelle recorded in the ground-layer. However, the greatest proportion of roosts occupied by individuals and small groups are found in the field- and shrub-layers, and those of larger groups in the shrub-layer alone.

2.5.2 Spring-flux

As with in winter, the BTHK Database holds records in all the layers during the spring-flux, and individual brown long-eared bats have been recorded in roosts with entrances in the ground-layer. However, the greater number of records of roosts occupied by individuals remain in the field- and shrub-layers though records of roosts in the canopy-layer are increasing in frequency.

2.5.3 Pregnancy

During the pregnancy period, although the BTHK Database has records of individuals, and of small and large groups in the field-layer, thus far there are no records of roosts in the ground-layer and overall, regardless of the tendency toward either individual or larger group roosting, the trend appears to suggest a preference for roosting higher-up.

2.5.4 Nursery

In the nursery period, although individuals appear to show no significant change from the pregnancy situation, smaller and larger groups are again more frequent and tend to be at their highest, and the larger groups are at their maximum height (although this may in fact reflect more tree climbing being done in this sunny and relatively windless season). Note there is a record of an individual Natterer's bat in the ground-layer.

2.5.5 Mating

In the mating period, whilst individuals do not significantly alter the broad preferences displayed in the pregnancy and nursery periods, the groups show a marked preference for lower roosts and are more frequent in the field- and shrub-layers. Nevertheless, the overall ranges for individuals and both small and large groups still span the ground- to the canopy-layer.

2.5.6 Autumn-flux

In the autumn-flux period individuals span all four height layers with brown long-eared bats again recorded in the ground-layer. As winter approaches, groups become smaller and although records still range high into the canopy, the general trend is for a descent into the shrub- and field-layers.

2.5.7 Summary

Figures 2.1, 2.2 and 2.3 illustrate the median, standard error and overall height ranges of tree-roost records occupied by one, two–four, and five or more bats in each of the six seasons.

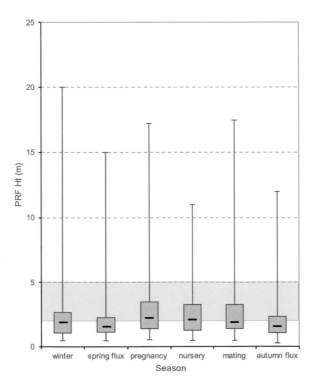

Figure 2.1 The median, standard error and range of heights occupied by individual bats during the six temporal periods in records held on the BTHK Database. The horizontal green band represents the shrub-layer.

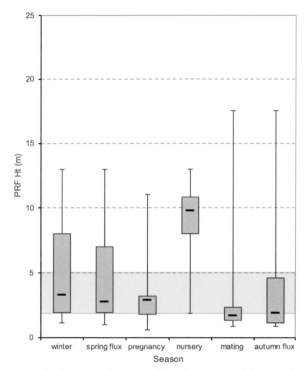

Figure 2.2 The median, standard error and range of heights occupied by two–four bats during the six temporal periods in records held on the BTHK Database. The horizontal green band represents the shrub-layer.

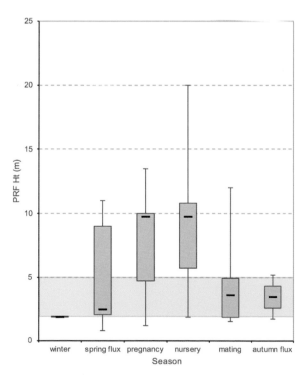

Figure 2.3 The median, standard error and range of heights occupied by five or more bats during the six temporal periods in records held on the BTHK Database. The horizontal green band represents the shrub-layer.

2.6 General tree-roost preferences

Different species favour different situations in which to roost.

In general terms, tree-roosting bats fall into three categories:

1. Void-roosting species, which are rarely encountered in situations where they are cramped, and comprise:
   a. Bechstein's bat;
   b. Daubenton's bat;
   c. The noctule; and
   d. The lesser horseshoe-bat.

2. Crevice-roosting species, which are rarely encountered in situations where they are in space, and comprise:
   a. The barbastelle;
   b. Alcathoe's bat;
   c. Brandt's bat;
   d. The whiskered bat;
   e. Nathusius' pipistrelle;
   f. The common pipistrelle; and
   g. The soprano pipistrelle.

Table 2.4 The broad roost preferences exhibited by the 14 tree-roosting bat species native to the British Isles

	General roost preferences	
	Void	Crevice
Barbastelle	–	✓
Bechstein's bat	✓	–
Alcathoe's bat	–	✓
Brandt's bat	–	✓
Daubenton's bat	✓	–
Natterer's bat	✓	✓
Whiskered bat	–	✓
Leisler's bat	✓	✓
Noctule	✓	–
Nathusius' pipistrelle	–	✓
Common pipistrelle	–	✓
Soprano pipistrelle	–	✓
Brown long-eared bat	✓	✓
Lesser horseshoe-bat	✓	–

3. Those that exploit both types and individuals and aggregations of the species may be encountered in voids and crevices. These comprise:

   a. Natterer's bat;

   b. Leisler's bat; and

   c. The brown long-eared bat.

These general roost preferences are illustrated at Table 2.4.

Of the overall 14 species, 10 exploit crevice-type PRFs and 7 show a significant preference for such features. To find the crevice-type PRF you have to be actively searching for them because, unlike woodpecker-holes, the entrances to the crevice-type PRFs are often camouflaged within the longitudinal fissures in bark and in cluttered situations behind epicormic growths[2] of twigs where you are unlikely to passively stumble upon them. However, it should be borne in mind that during the mating period, pairs of bats may occupy situations that might be considered atypical at any other time of year.

2.6.1 Summing-up

In summary, there are 14 bat species that variously occupy trees in six broad habitat types, during six temporal periods in four spatial zones, and favouring two distinct situations in which to rest: voids and crevices.

The tables and figures presented in this chapter were designed for practical use. They have been thoroughly tested to ensure that when they are combined with the PRF descriptions in the following chapter, they provide an effective foundation for a desk-study and subsequent truthing framework, and in all cases illustrate the situations that are 'more-likely-than-not'.

---

2 Juvenile shoots originating from live tissues of current sapwood just under the bark (Watson 2006). The leaves on these twigs are often significantly larger than those in canopy foliage and obstruct inspection of the bark behind. However, by understanding that they often result from damage, a positive-spin may be applied to the situation and these epicormic twigs with big leaves on the sides of limbs can be used as field-signs for hazard-beams and shearing-cracks (see Chapter 3 for descriptions).

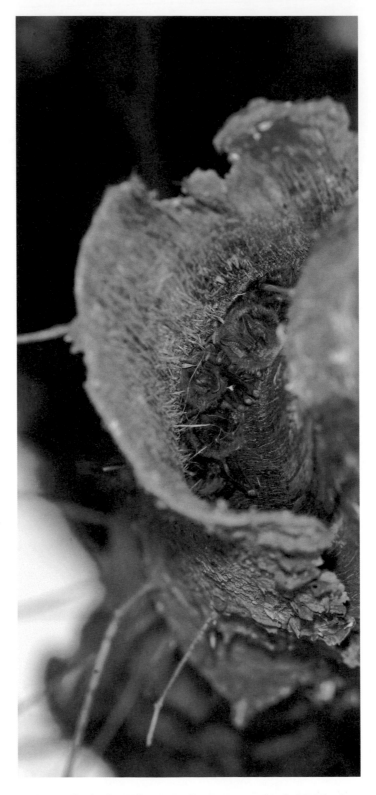

The Grail: a barbastelle maternity group roosting behind bark
(photo René Janssen).

CHAPTER 3

Potential Roost Features (PRFs)

In this chapter	
Introduction	The primary mechanisms of PRF formation
The Disease & Decay PRFs	The agents of disease and decay and the PRFs that result from it
The Damage PRFs	The agents of damage and the PRFs that result from it
The Association PRFs	The agents of association and the PRFs it encompasses

3.1 Introduction

Bats do not make the features they roost in but rely upon the chance presence of three mechanisms to provide them with shelter. These comprise:

1. Disease & Decay.
2. Damage.
3. Associations.

These three PRF forms result in a variety of different features but they can be placed in the two roost-preference categories that were identified at the close of the preceding chapter, i.e. voids and crevices.

The three PRF forms comprise:

» **Disease & Decay**, which encompasses:
   a. Woodpecker-holes;
   b. Squirrel-holes;
   c. Knot-holes;
   d. Pruning-cuts;
   e. Tear-outs;
   f. Wounds;
   g. Cankers;
   h. Compression-forks; and
   i. Butt-rots.
» **Damage**, which encompasses:
   a. Lightning-strikes;
   b. Hazard-beams;
   c. Subsidence-cracks;

    d.   Shearing-cracks;

    e.   Transverse-snaps;

    f.   Welds;

    g.   Lifting bark;

    h.   Desiccation-fissures; and

    i.   Frost-cracks.

» **Association**, which encompasses:

    a.   Fluting; and

    b.   Ivy.

The various mechanisms are neatly compartmentalised above, but in fact the situation is complicated in that some occur in combination; for example, a knot-hole might be opened by a woodpecker, and a wound may be prevented from healing by a squirrel. Furthermore, in some situations the PRF forms cross boundaries, i.e. it is not uncommon to find subsidence-cracks that are the result of a butt-rot but were only opened to the bats by a grey squirrel. All the classifications do is identify the component mechanisms in what might be a much more complicated process.

Each of the 20 PRF types is described in the following subsections. The descriptions are tabulated summaries of records held in the BTHK Database and detailed descriptions are provided in the *Bat Tree Habitat Key* (Andrews *et al.* 2016), both of which may be accessed at www.battreehabitatkey.com. The information is simplified here in order that it may be put to practical use in the performance of a desk-study, and in the field for identification.

Where data exist on the BTHK Database, the diameter at breast height (DBH[1]) ranges of roost trees are identified for each period.

Using the BTHK Database and referring to the species accounts within the *Bat Tree Habitat Key*, the bat species for which encounter is 'more-likely-than-not' in the individual PRF are identified as follows:

» **Winter:** occupation by any number of bats during the Winter period (i.e. January and February).

» **Transitory:** occupation by 1–3 bats during the Pregnancy, Nursery and both Flux periods (i.e. March–August and November/December).

» **Maternity:** occupation by a maternity colony within the Pregnancy and Nursery periods (i.e. May–August).

» **Mating:** occupation by any number of bats during the Mating period (i.e. September and October).

It should be noted that the DBH ranges are for all roosts occupied within each period. Therefore, the ranges identified for the Maternity period comprise both roosts occupied by colonies and by individuals (i.e. transitory occupation). Furthermore, the summaries are not exhaustive but apply the principle of 'more-likely-than-not' (i.e. "*reasonable likelihood*"). In particular, the tree species set out are not comprehensive, but list those species that are 'more-likely-than-not' to hold that PRF type and for which sufficient roost records exist to suggest they are consistently occupied in the context identified.

---

1 DBH is the standard measurement used to calculate the volume of wood in a single tree and a plantation crop. The diameter is taken using a tape or callipers on the stem (trunk) or bole at approximately 1.4 m (to be fully accurate the height is 4′ 6″, or 137 cm). For the purpose of this text, DBH is used to compare the sizes of standing trees against data held on the BTHK Database.

Each PRF form closes with two summary figures. The first identifies which bat species has been recorded occupying the individual PRF types and when. The second figure identifies which PRF are 'more-likely-than-not' to occur in each habitat type. These figures are designed for practical use in a desk-study.

3.2 The Disease & Decay PRFs

3.2.1 Introduction

The principal agents of disease and decay are:

» **The white-rots**, which mainly affect broadleaves and are most prevalent in the crown. The white rots result in softening of the heartwood into damp humus.

» **The brown-rots**, which mainly affect conifers, although the broadleaved heartwood species also have associated brown rots, i.e. chicken of the woods *Laetiporus sulphurous* is a brown rot and infects oaks, sweet chestnut, black locust as well as yew. The brown-rots are most prevalent in the butt, as they tend to be transmitted via the root system, and result in cubical cracking of the heartwood into a desiccated system of fissures, and ultimately into dust.

From a bat-roost perspective, white-rots are typically damp and most often occur from the shrub-layer into the canopy, but brown-rots are always very dry and typically occur from the ground into the shrub-layer.

Accounts of each of the Disease & Decay PRFs are provided in Table 3.1 to Table 3.9.

Table 3.1 Woodpecker-holes

CAUSED BY	Green woodpecker *Picus viridis*, great spotted woodpecker *Dendrocopos major* and lesser spotted woodpecker *Dendrocopos minor*		
DESCRIPTION	A 'gourd-shaped' void below a 3–7 cm diameter circular entrance. Where a void exists above the entrance, it may have either a tall spire or a shallowly domed apex. Woodpecker-holes most frequently occur on the stem and near-vertical limbs, but may also occur on the underside of a leaning limb. Where the void extends above the entrance, the bats will typically (but not always) be in the apex. Where the void extends downward, an aggregation will be in the base looking up. If they are noctules the entrance is often flanked by one or two bats facing horizontally. If a noctule mating-roost, the male will be on the opposite wall facing the entrance and the females will be behind him, either above or below		

TYPE	Crevice	—	
	Void	✓	

APEX SHAPE	Dome	✓	
	Peak/Wedge	—	
	Spire	✓	

HABITAT	Continuous canopy	Woodland	✓
		Plantation	✓
		Orchard	—
	Linear	Hedge / tree-line	✓
		Shelter-belt	—
		Riparian fringe	✓
	Loose canopy	Parkland	✓
		Churchyard & garden	—

TOPOGRAPHY	Level	✓	
	Sloping	✓	

SITUATION	Central	✓	
	Corridor	✓	
	Edge	✓	

ROOST TREE SPECIES	Alder, ash, hybrid black poplar, sessile oak, pedunculate oak, goat/grey willow, white/crack willow and dead wood on Scots and Corsican pines			

DBH RANGE OF ROOST TREES OCCUPIED IN EACH PERIOD	Jan/Feb	Mar/Apr & Nov/Dec	May–Aug	Sep/Oct
	25–93.6 cm	25–98.4 cm	25–150 cm	25–93.6 cm

HEIGHT RANGE OCCUPIED IN EACH PERIOD	Jan/Feb	Mar/Apr & Nov/Dec	May–Aug	Sep/Oct
	2.67–8.38 m	2.1–7.5 m	1.62–14 m	2.1–8.38 m

BAT SPECIES	ROOST PURPOSE			
	Winter	Transitory	Maternity	Mating
Bechstein's bat	—	—	✓	—
Daubenton's bat	—	—	✓	—
Natterer's bat	—	✓	✓	—
Leisler's bat	—	—	✓	—
Noctule	✓	✓	✓	✓
Soprano pipistrelle	—	✓	—	—
Brown long-eared bat	—	✓	✓	✓

Woodpecker-holes: Top to bottom: Natterer's bat; noctules (note group is below entrance); and brown long-eared bats.

33

Table 3.2 Squirrel-holes

CAUSED BY	Grey squirrels *Sciurus carolinensis*			
DESCRIPTION	A cylindrical void above a 6–8 cm diameter circular entrance with either a dome or spire apex. Grey squirrels may rest on the base of a PRF during the day but sleep well above the entrance (i.e. >50 cm up). Squirrel-holes may therefore be separated from woodpecker-holes by the fact that the void does not extend much (if at all) below the entrance, which very often has a 'cowled' appearance due to the squirrels gnawing the basal rim significantly more diligently than the upper rim. Additional cues are the association of the entrance with another mechanism (most commonly a knot-hole or wound that has been opened and maintained open by gnawing), the scent of hay is often detectable in the entrance, and 2–3 cm white/grey tail hairs may get stuck to the entrance rim. Obviously if the animal is at home, there will be a large angry ball of grey fur 'churring' in the apex			
TYPE	Crevice	—		
	Void	✓		
APEX SHAPE	Dome	✓		
	Peak/Wedge	—		
	Spire	✓		
HABITAT	Continuous canopy	Woodland	✓	
		Plantation	—	
		Orchard	—	
	Linear	Hedge / tree-line	—	
		Shelter-belt	—	
		Riparian fringe	—	
	Loose canopy	Parkland	—	
		Churchyard & garden	—	
TOPOGRAPHY	Level	✓		
	Sloping	✓		
SITUATION	Central	✓		
	Corridor	✓		
	Edge	✓		
ROOST TREE SPECIES	Ash and sessile oak			

DBH RANGE OF ROOST TREES OCCUPIED IN EACH PERIOD	Jan/Feb	Mar/Apr & Nov/Dec	May–Aug	Sep/Oct
	31 cm (data-deficient)	31–49.5 cm	31–49.5 cm	31 cm (data-deficient)

HEIGHT RANGE OCCUPIED IN EACH PERIOD	Jan/Feb	Mar/Apr & Nov/Dec	May–Aug	Sep/Oct
	2.65 m (data-deficient)	2.65-4.79 m	2.65-4.79 m	2.65 m (data-deficient)

BAT SPECIES	ROOST PURPOSE			
	Winter	Transitory	Maternity	Mating
Natterer's bat	—	✓	—	—
Noctule	—	—	✓	—
Brown long-eared bat	—	—	✓	—
Unknown	✓	—	—	—

Squirrel-holes: Top to bottom: Natterer's bats; noctules; and, brown long-eared bat.

Table 3.3 Knot-holes

CAUSED BY	Natural limb-loss brought about by competition for light			
DESCRIPTION	A cylindrical or spherical void. Knot-holes occur on stems and larger limbs with the entrance at any orientation. In terms of height, knot-holes are typical of the shrub and canopy-layers, but do occasionally crop up in the field-layer. Knot-hole roosts are often spherical with the entrance mid-way up the side wall. Where the void extends above the entrance the bats will be in the apex. However, knot-holes typically have a shallowly domed apex and basal cup; rather like a small football. The bats may be on the back wall, opposite the entrance and even beneath it			
TYPE	Crevice	—		
	Void	✓		
APEX SHAPE	Dome	✓		
	Peak/Wedge	—		
	Spire	—		
HABITAT	Continuous canopy	Woodland	✓	
		Plantation	✓ (broadleaved)	
		Orchard	—	
	Linear	Hedge / tree-line	✓	
		Shelter-belt	—	
		Riparian fringe	—	
	Loose canopy	Parkland	✓	
		Churchyard & garden	—	
TOPOGRAPHY	Level	✓		
	Sloping	✓		
SITUATION	Central	✓		
	Corridor	✓		
	Edge	✓		
ROOST TREE SPECIES	Field maple, Norway maple, silver birch, beech, ash, domestic apple, hybrid black poplar, sessile oak and pedunculate oak			

DBH RANGE OF ROOST TREES OCCUPIED IN EACH PERIOD	Jan/Feb	Mar/Apr & Nov/Dec	May–Aug	Sep/Oct
	31.9–43.2 cm	31.9–116 cm	31.5–120.3 cm	27–180 cm

HEIGHT RANGE OCCUPIED IN EACH PERIOD	Jan/Feb	Mar/Apr & Nov/Dec	May–Aug	Sep/Oct
	0.92–2.18 m	2.18–9 m	1.43–13 m	1.43–6 m

BAT SPECIES	ROOST PURPOSE			
	Winter	Transitory	Maternity	Mating
Bechstein's bat	—	✓	✓	✓
Daubenton's bat	—	✓	✓	✓
Natterer's bat	✓	—	—	✓
Leisler's bat	—	✓	✓	—
Noctule	—	✓	✓	✓
Nathusius' pipistrelle	—	✓	—	—
Common pipistrelle	—	✓	—	✓
Soprano pipistrelle	—	✓	—	—
Brown long-eared bat	✓	—	✓	—

Knot-holes: Top to bottom: brown long-eared bat; common pipistrelle; and, soprano pipistrelle.

Table 3.4 Pruning-cuts

CAUSED BY	The removal of a limb by a saw, typically too close to the stem and thereby removing the branch-collar, or by a correct cut administered to a tree that is moribund or growing in an environment that is inhospitable and retards the healing process			
DESCRIPTION	A spherical void or an arched crevice. Typically, on the stem or vertical limbs. By far the greater proportion confined to shrub-layer limbs that are removed to allow unobstructed traffic. The bat-roost position is typically above the entrance or, as with the knot-hole description; opposite the entrance			

TYPE	Crevice	✓		
	Void	✓		
APEX SHAPE	Dome	✓		
	Peak/Wedge	—		
	Spire	—		

HABITAT	Continuous canopy	Woodland	—
		Plantation	—
		Orchard	✓
	Linear	Hedge / tree-line	—
		Shelter-belt	—
		Riparian fringe	—
	Loose canopy	Parkland	✓
		Churchyard & garden	—

TOPOGRAPHY	Level	✓
	Sloping	✓
SITUATION	Central	✓
	Corridor	✓
	Edge	✓

ROOST TREE SPECIES	Ash and pedunculate oak			

DBH RANGE OF ROOST TREES OCCUPIED IN EACH PERIOD	Jan/Feb	Mar/Apr & Nov/Dec	May–Aug	Sep/Oct
	—	75 cm (data-deficient)	75–99.6 cm	—

HEIGHT RANGE OCCUPIED IN EACH PERIOD	Jan/Feb	Mar/Apr & Nov/Dec	May–Aug	Sep/Oct
	—	5–6.5 m	5-6.8 m	—

BAT SPECIES	ROOST PURPOSE			
	Winter	Transitory	Maternity	Mating
Daubenton's bat	—	—	✓ (void)	—
Natterer's bat	—	—	✓ (void)	—
Leisler's bat	—	✓ (crevice)	—	—
Noctule	—	—	✓ (void)	—
Common pipistrelle	—	—	✓ (void)	—

Pruning cuts: Top to bottom: noctules; unknown; and, unknown.

Table 3.5 Tear-outs

CAUSED BY	The tearing down and out of a limb, due to snow-weight, impact from an adjacent tree falling, and (less commonly) wind		
DESCRIPTION	A narrow void-type PRF, typically with an elongated spire apex. A feature of the stem and, very occasionally, a vertical limb. The bat-roost position is most commonly well above the entrance		
TYPE	Crevice	—	
	Void	✓	
APEX SHAPE	Dome	✓	
	Peak/Wedge	—	
	Spire	✓	
HABITAT	Continuous canopy	Woodland	✓
		Plantation	—
		Orchard	—
	Linear	Hedge / tree-line	—
		Shelter-belt	—
		Riparian fringe	—
	Loose canopy	Parkland	✓
		Churchyard & garden	✓
TOPOGRAPHY	Level	✓	
	Sloping	✓	
SITUATION	Central	✓	
	Corridor	✓	
	Edge	✓	
ROOST TREE SPECIES	Field maple, sycamore, horse chestnut, alder, silver birch, downy birch, ash, sweetgum, wild cherry, sessile oak, pedunculate oak and goat/grey willow		

DBH RANGE OF ROOST TREES OCCUPIED IN EACH PERIOD	Jan/Feb	Mar/Apr & Nov/Dec	May–Aug	Sep/Oct
	19–118 cm	15–118 cm	24.8–118 cm	28–118 cm

HEIGHT RANGE OCCUPIED IN EACH PERIOD	Jan/Feb	Mar/Apr & Nov/Dec	May–Aug	Sep/Oct
	0.5–13 m	1.4–17.5 m	1.4–13 m	0.5–17.5 m

BAT SPECIES	ROOST PURPOSE			
	Winter	Transitory	Maternity	Mating
Barbastelle	—	—	✓	—
Bechstein's bat	—	✓	✓	✓
Daubenton's bat	—	✓	✓	✓
Natterer's bat	—	✓	✓	✓
Leisler's bat	—	✓	✓	✓
Noctule	✓	✓	—	✓
Nathusius' pipistrelle	—	✓	—	—
Common pipistrelle	—	✓	—	—
Soprano pipistrelle	✓	✓	—	✓
Brown long-eared bat	✓	✓	—	✓

Tear-outs: Top to bottom: Daubenton's bats; unknown; and, common pipistrelle.

Table 3.6 Wounds

CAUSED BY	Rot following the loss of bark and cambium due to: abrasion from a falling stem; a rubbing action; an impact; or bark-stripping by squirrels			
DESCRIPTION	Wounds typically resemble a dug-out canoe. Sometimes the wound may appear to be sealed by callus against a core of exposed heartwood, but close-inspection reveals a narrow opening at the wound apex caused by heartwood decay. Stem wounds are typical of the field-layer, but on limbs they occur into the high canopy and may be invisible from the ground			

TYPE	Crevice	—		
	Void	✓		

APEX SHAPE	Dome	✓		
	Peak/Wedge	✓ (see Jim-Gem description)		
	Spire	✓ (see Jim-Gem description)		

HABITAT	**Continuous canopy**	Woodland	✓
		Plantation	✓
		Orchard	✓
	Linear	Hedge / tree-line	✓
		Shelter-belt	—
		Riparian fringe	—
	Loose canopy	Parkland	✓
		Churchyard & garden	✓

TOPOGRAPHY	Level	✓		
	Sloping	✓		

SITUATION	Central	✓		
	Corridor	✓		
	Edge	✓		

ROOST TREE SPECIES	Field maple, sycamore, alder, sweet chestnut, hazel, beech, ash, domestic apple, sessile oak, pedunculate oak, goat/grey willow, white/crack willow, rowan, elm and Scots pine			

DBH RANGE OF ROOST TREES OCCUPIED IN EACH PERIOD	Jan/Feb	Mar/Apr & Nov/Dec	May–Aug	Sep/Oct
	8.7–76.8 cm	10.1–98.4 cm	10.1–89.7 cm	10.1–120 cm

HEIGHT RANGE OCCUPIED IN EACH PERIOD	Jan/Feb	Mar/Apr & Nov/Dec	May–Aug	Sep/Oct
	0.68–20 m	0.6–8 m	0.5–20 m	0.6–7.6 m

BAT SPECIES	ROOST PURPOSE			
	Winter	Transitory	Maternity	Mating
Barbastelle	—	✓	—	✓
Bechstein's bat	—	✓	✓	—
Daubenton's bat	—	✓	✓	✓
Whiskered bat	—	✓	—	—
Natterer's bat	✓	✓	✓	✓
Leisler's bat	✓	✓	✓	✓
Noctule	✓	✓	✓	—
Common pipistrelle	✓	✓	—	✓
Soprano pipistrelle	—	✓	✓	✓
Brown long-eared bat	✓	✓	✓	✓
Lesser horseshoe-bat	—	✓	—	—

Wounds: Top to bottom: brown long-eared bats; Natterer's bat; and, brown long-eared bat.

Table 3.7 Cankers

CAUSED BY	A bacterial infection			
DESCRIPTION	A pitted, blistered or sunken look to the entrance. On sycamore, silver birch, ash and wild cherry it typically looks scabby, and on horse chestnut and beech sunken and drawn. Most commonly cankers offer a cylindrical void, but not always. Cankers are typically orientated vertically and can occur from the butt into the canopy. The bat roost position is typically above the entrance in void situations, but in crevices the bats may be above, below and even to one side			
TYPE	**Crevice**	✓		
	Void	✓		
APEX SHAPE	**Dome**	—		
	Peak/Wedge	✓		
	Spire	✓		
HABITAT	**Continuous canopy**	Woodland	✓	
		Plantation	—	
		Orchard	—	
	Linear	Hedge / tree-line	✓	
		Shelter-belt	—	
		Riparian fringe	—	
	Loose canopy	Parkland	—	
		Churchyard & garden	—	
TOPOGRAPHY	**Level**	✓		
	Sloping	✓		
SITUATION	**Central**	✓		
	Corridor	✓		
	Edge	✓		
ROOST TREE SPECIES	Sycamore, horse chestnut, silver birch, beech, ash and wild cherry			

DBH RANGE OF ROOST TREES OCCUPIED IN EACH PERIOD	Jan/Feb	Mar/Apr & Nov/Dec	May–Aug	Sep/Oct
	10–15.4 cm	10–66 cm	10–76.9 cm	20.5–28 cm

HEIGHT RANGE OCCUPIED IN EACH PERIOD	Jan/Feb	Mar/Apr & Nov/Dec	May–Aug	Sep/Oct
	1.3–2.86 m	1.3–4.2 m	1.3–17.2 m	2.2–2.5 m

BAT SPECIES	ROOST PURPOSE			
	Winter	**Transitory**	**Maternity**	**Mating**
Daubenton's bat	—	✓	✓	✓
Natterer's bat	✓	✓	✓	✓
Leisler's bat	✓	✓	—	—
Noctule	✓	—	✓	—
Nathusius' pipistrelle	—	—	—	✓
Common pipistrelle	✓	✓	—	—
Soprano pipistrelle	—	✓	—	—
Brown long-eared bat	✓	✓	—	✓

Cankers: Top to bottom: Natterer's bats; unknown; and Daubenton's bats.

Table 3.8 Compression-forks

CAUSED BY	A growth deformity. Coniferous trees typically have one apical bud, known as 'the leader'. If you think of a Christmas tree, the leader is where you put the star. If this bud is removed, by an insect, bird or mammal, the tree will respond by putting on two. Some broadleaves, notably beech and cherry (amongst others), have a predisposition for dividing in parallel pairs of leaders, resulting in two limbs growing in parallel, which places a stress on the fork beneath. The bases of the limbs push against each other as they put on successive growth, and also pull against each other as they are blown in the wind. Over time a split may occur in the stem beneath, or damp litter may build-up between them, and this may result in an infection by a pathogen and hollowing extending downward into the trunk beneath			
DESCRIPTION	A cylindrical void in the bole accessed from an elliptic entrance at the top between the co-dominant stems. On conifers compression-forks are always in the stem. On a broadleaved tree there may be multiple compression forks into the canopy. Where the void is accessed from between the leaders, the bats will obviously be below, but where it is accessed from a split in the stem below the fork, they may be on either side of the entrance			

TYPE	Crevice	—		
	Void	✓		

APEX SHAPE	Dome	N/A		
	Peak/Wedge	N/A		
	Spire	N/A		

HABITAT	**Continuous canopy**	Woodland	—
		Plantation	✓
		Orchard	—
	Linear	Hedge / tree-line	—
		Shelter-belt	—
		Riparian fringe	—
	Loose canopy	Parkland	✓
		Churchyard & garden	—

TOPOGRAPHY	Level	✓
	Sloping	—

SITUATION	Central	✓
	Corridor	✓
	Edge	✓

ROOST TREE SPECIES	Beech, European larch, Corsican pine, Scots pine, Douglas fir and yew			

DBH RANGE OF ROOST TREES OCCUPIED IN EACH PERIOD	Jan/Feb	Mar/Apr & Nov/Dec	May–Aug	Sep/Oct
	47–118 cm	47–118 cm	89.2 cm (data-deficient)	—

HEIGHT RANGE OCCUPIED IN EACH PERIOD	Jan/Feb	Mar/Apr & Nov/Dec	May–Aug	Sep/Oct
	1.6–2.1 m	1.6–2.1 m	4.48 m (data-deficient)	—

BAT SPECIES	ROOST PURPOSE			
	Winter	Transitory	Maternity	Mating
Natterer's bat	—	—	✓	—
Brown long-eared bat	—	✓	✓	—

Compression-forks: Top row: Natterer's bats; middle row: context example; and bottom row: Natterer's bats (bottom three photos: K. Cohen).

Table 3.9 Butt-rots

CAUSED BY	Brown-rot entering from the roots		
DESCRIPTION	In broadleaves, the result may be a 'chimney-like' cylindrical void with a dome or spire apex. In conifers, they are more typically elongated crevices. Always vertically orientated with the entrance at the base of the stem. In broadleaves, the bats are found in the apex. In conifers, the bats may be in the core or between the cambium and heartwood above, below and to one side of the entrance		

| **TYPE** | Crevice | ✓ (conifers) | |
| | Void | ✓ (broadleaves) | |

APEX SHAPE	Dome	✓	
	Peak/Wedge	—	
	Spire	✓	

HABITAT	Continuous canopy	Woodland	✓
		Plantation	✓
		Orchard	—
	Linear	Hedge / tree-line	—
		Shelter-belt	—
		Riparian fringe	—
	Loose canopy	Parkland	✓
		Churchyard & garden	—

| **TOPOGRAPHY** | Level | ✓ | |
| | Sloping | ✓ | |

SITUATION	Central	✓	
	Corridor	✓	
	Edge	✓	

ROOST TREE SPECIES	Beech, London plane, Turkey oak, limes, Scots pine and Douglas fir			
DBH RANGE OF ROOST TREES OCCUPIED IN EACH PERIOD	Jan/Feb	Mar/Apr & Nov/Dec	May–Aug	Sep/Oct
	178 cm (data-deficient)	62–258.2 cm	64–121.3 cm	258.2 cm (data-deficient)
HEIGHT RANGE OCCUPIED IN EACH PERIOD	Jan/Feb	Mar/Apr & Nov/Dec	May–Aug	Sep/Oct
	1.94 m (data-deficient)	0.86–5.1 m	0.57–0.6 m	0.86 m (data-deficient)
BAT SPECIES	ROOST PURPOSE			
	Winter	Transitory	Maternity	Mating
Daubenton's bat	—	✓	—	—
Common pipistrelle	—	✓	—	—
Soprano pipistrelle	✓	—	—	—
Lesser horseshoe-bat	—	✓	—	—

Butt-rots: Top to bottom: lesser horseshoe-bats (top middle and right photos: J. Shipman); lesser horseshoe-bats; and lesser horseshoe-bats.

3.2.2 Jim-Gem

The Jim-Gem was invented by an American, Jim Craig. Essentially it is a large syringe, about 1 m long and 10 cm in diameter, which is used to inject herbicide into suckering trees that are not killed by felling. It has a sharp, chisel-like blade that is swung into the tree and the pump lever pushes in herbicide from the cylindrical reservoir. By girdling the stem the tree above is killed and the herbicide thereby travels down and kills the root system. In America, it has been shown to be effective for aspen *Populus tremula* and various birch species *Betula* spp.

This novel invention was used in a failed experiment which attempted to eradicate lime *Tilia* sp. and sweet chestnut *Castanea sativa* stools in unprofitable coppice woodland (O. Rackham, personal communication, August 2012). Although its application was unsuccessful in killing the coppice stools, from the perspective of creating roost habitat it was a resounding success and in some instances has resulted in a high-density abundance of living stems which have been left as standing dug-out canoes. Three bat species have been encountered roosting in these features: Natterer's bat *Myotis nattereri*, the common pipistrelle *Pipistrellus pipistrellus* and the brown long-eared bat *Plecotus auritus*.

3.2.3 Summary

Table 3.10 illustrates which features comprise crevices, voids or both, and summarises the bat species that have been recorded exploiting the individual Disease & Decay PRFs. Importantly, the table also identifies the PRF for which there are insufficient grounds to suggest that exploitation by an individual bat species might be 'more-likely-than-not'.

Table 3.10 An illustration of which Disease & Decay PRFs comprise crevices, voids or both, and which bat species have been recorded roosting in the individual PRF types and for what purpose

DISEASE & DECAY PRF SUMMARY	CREVICE	VOID	Barbastelle	Bechstein's bat	Alcathoe's bat	Brandt's bat	Daubenton's bat	Whiskered bat	Natterer's bat	Leisler's bat	Noctule	Nathusius' pipistrelle	Common pipistrelle	Soprano pipistrelle	Brown long-eared bat	Lesser horseshoe-bat
Woodpecker-holes	–	✓	–	PN	–	–	PN	–	PN T	PN	PN T M W	–	–	T	PN T M	–
Squirrel-holes	–	✓	–	–	–	–	–	–	T	–	PN	–	–	–	PN	–
Knot-holes	–	✓	–	PN T M	–	–	PN T M	–	M W	PN T	PN T M	T	T M	T	M W	–
Pruning-cuts	✓	✓	–	–	–	–	PN	–	PN	T	PN	–	PN	–	–	–
Tear-outs	–	✓	PN	PN T M	–	–	PN T M	–	PN T M	PN T	T M W	T	T	T M W	T M W	–
Wounds	–	✓	T M	PN T	–	–	PN T M	I	PN T M W	PN T M W	PN T W	–	T M W	PN T M	PN T W M	T
Cankers	✓	✓	–	–	–	–	PN T M	–	PN T M W	T W	PN W	M	T W	T	T M W	–
Compression-forks	–	✓	–	–	–	–	–	–	PN	–	–	–	–	–	PN T	–
Butt-rots	✓	✓	–	–	–	–	T	–	–	–	–	–	–	T	W	T

PN, occupation by a maternity colony within the Pregnancy and Nursery periods; T, occupation by 1–3 bats during the Pregnancy, Nursery and both Flux periods; M, occupation by any number of bats during the Mating period; W, occupation by any number of bats during the Winter period.

The Disease & Decay PRFs comprise a significantly greater number of void-type PRF.

Table 3.11 summarises the habitat types in which the individual Disease & Decay PRFs have been recorded and exploited by roosting bats.

Table 3.11 The habitat types in which the individual Disease & Decay PRFs have been recorded and exploited by roosting bats

CANOPY COVER	HABITAT TYPE	PRF TYPE								
		Woodpecker-holes	Squirrel-holes	Knot-holes	Pruning-cuts	Tear-outs	Wounds	Cankers	Compression-forks	Butt-rots
CONTINUOUS CANOPY	Woodland	✓	✓	✓	—	✓	✓	✓	—	✓
	Plantation	✓	—	✓	—	—	✓	—	✓	✓
	Orchard	—	—	—	✓	—	✓	—	—	—
LINEAR CANOPY	Hedge/tree-line	✓	—	✓	—	—	—	✓	—	—
	Shelter-belt	—	—	—	—	—	—	—	—	—
	Riparian-fringe	✓	—	—	—	—	—	—	—	—
LOOSE CANOPY	Parkland	✓	—	✓	✓	✓	—	—	✓	—
	Churchyard & garden	—	—	—	—	—	✓	—	—	—

3.3 The Damage PRFs

Damage may occur as the result of a plethora of environmental and physical influences, including: sudden changes in temperature; root compaction; bad tree surgery; weight of snow; gnawing by squirrels, deer, livestock and invertebrates; impacts (typically an adjacent tree falling); wind; compression (i.e. the weight of the tree bearing down on itself); and frost. However, the first five of these mechanisms require pathogens for them to progress into PRFs, whereas the last four just require the environment and the tree's attempts to mitigate against it. The important damage mechanisms therefore comprise:

1. Impact.
2. Wind.
3. Compression.
4. Frost.

These four mechanisms result in splitting, over which the tree will gradually put on woundwood each year, slowly and incrementally sealing across the wound which is ultimately covered by a curtain of tissue. Over time, this results in a nice darkened crevice-type PRF, with an elongated narrow entrance that predators cannot get through; ideal as a secure roost environment and almost always with a roost position above the entrance.

The healing process described above results in a characteristic that is particular to the Damage PRF, namely 'ram's-horns'.

3.3.1 Ram's-horns

Ram's-horns are the consequence of the rapid growth of woundwood following a longitudinal split, impact wound or bark-stripping (Lonsdale 1999). They are so called because of the appearance of the woundwood in cross-section. Figure 3.1 illustrates the characteristics of the feature.

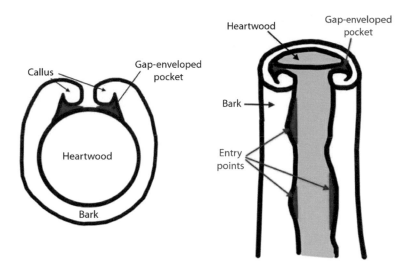

Figure 3.1 Ram's-horns shown in cross-section (left) and longitudinally (right).

The feature forms when new growth turns over into itself and often, whilst the original gap is reduced, it may not entirely close the initial split even where the horns meet, but often leaves irregular elliptical openings. However, it is more common for the growth to roll into itself leaving a tubular feature inside. As a result, ram's-horns are often a PRF in their own right, comprising a secondary feature below the primary and most obvious PRF. Care must be taken not to overlook them as they are exploited by barbastelles *Barbastella barbastellus* (a particular favourite of individual bats), common pipistrelles and brown long-eared bats.

Accounts of each of the Damage PRFs are provided in Table 3.12 to Table 3.20.

Table 3.12 Lightning-strikes

CAUSED BY	The sudden and violent expansion of sap, which has been instantly turned to steam by the passage of electricity through the trees tissue, blasting a strip of bark off and away from the stem			
DESCRIPTION	Damage may take two forms: scarring/cracking, and complete rending of the tree. In the latter case the conspicuous fissure often extends from the tip of the leader to the base of the tree. In extreme cases the tree is killed, but in others a strip of bark and young wood with more or less parallel sides is split along the entire length. The PRF has a vertical or corkscrewing orientation up the stem. Occasionally there are two splits opposite each other on the stem. The bat roost position is in the gap formed inside the cambium-layer. As a result, lightning-strikes are the trickiest PRF there is! The feature offers huge crevices and ram's-horns from the butt and into the highest point of the crown, although the strike can also hit obliquely in the lower canopy. The roost position may be to the side, in and down, or in and up, and there may be corners to negotiate			

TYPE	Crevice	✓		
	Void	—		

APEX SHAPE	Dome	✓		
	Peak/Wedge	✓		
	Spire	—		

HABITAT	Continuous canopy	Woodland	✓
		Plantation	—
		Orchard	—
	Linear	Hedge / tree-line	—
		Shelter-belt	✓
		Riparian fringe	—
	Loose canopy	Parkland	✓
		Churchyard & garden	—

TOPOGRAPHY	Level	✓
	Sloping	✓

SITUATION	Central	✓
	Corridor	✓
	Edge	✓

ROOST TREE SPECIES	Beech, sessile oak and pedunculate oak			

DBH RANGE OF ROOST TREES OCCUPIED IN EACH PERIOD	Jan/Feb	Mar/Apr & Nov/Dec	May–Aug	Sep/Oct
	31.1 cm (data-deficient)	49.7 cm (data-deficient)	31.1–159.6 cm	31.1–159.6 cm

HEIGHT RANGE OCCUPIED IN EACH PERIOD	Jan/Feb	Mar/Apr & Nov/Dec	May–Aug	Sep/Oct
	1.89 m (data-deficient)	5.96 m (data-deficient)	1.89–10.8 m	1.89–10.8 m

BAT SPECIES	ROOST PURPOSE			
	Winter	Transitory	Maternity	Mating
Barbastelle	—	—	✓	—
Daubenton's bat	—	—	✓	—
Natterer's bat	—	✓	—	—
Leisler's bat	—	—	—	✓
Brown long-eared bat	—	✓	✓	—

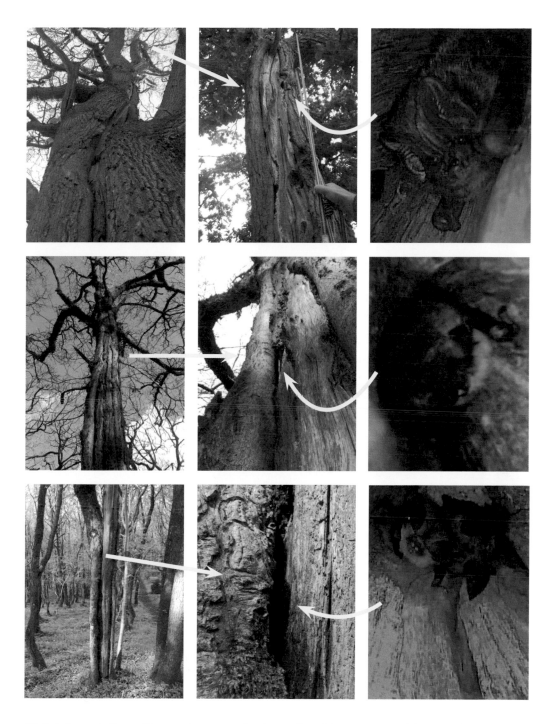

Lightning-strikes: Top to bottom: barbastelles; Natterer's and brown long-eared bats; and brown long-eared bats.

Table 3.13 Hazard-beams

CAUSED BY	Wind turbulence violently twisting and bending the limb at the same time			
DESCRIPTION	A diametral split across the ray-wood, spanning the stem or limb completely, and resulting in an elongated crevice, widest in the middle and extending up and down into a gradually narrowing peak. Accessed from either side via entrances that are directly opposite each other. Occurs on both stems and limbs (but far more common on limbs) and provides a roost position above the entrance			
TYPE	Crevice	✓		
	Void	✓ in the lower section (the apex is always a crevice)		
APEX SHAPE	Dome	—		
	Peak/Wedge	✓		
	Spire	—		
HABITAT	Continuous canopy	Woodland	✓	
		Plantation	—	
		Orchard	—	
	Linear	Hedge / tree-line	✓	
		Shelter-belt	—	
		Riparian fringe	✓	
	Loose canopy	Parkland	✓	
		Churchyard & garden	✓	
TOPOGRAPHY	Level	✓		
	Sloping	✓		
SITUATION	Central	—		
	Corridor	✓		
	Edge	✓		
ROOST TREE SPECIES	Horse chestnut, beech, sweet chestnut, ash, sessile oak, pedunculate oak, black locust, Monterey cypress, European larch, Scots pine (latter two in open-grown conditions only) and yew			

DBH RANGE OF ROOST TREES OCCUPIED IN EACH PERIOD	Jan/Feb	Mar/Apr & Nov/Dec	May–Aug	Sep/Oct
	46–149.5 cm	46–149.5 cm	30.1–451.2 cm	0.9–150 cm

HEIGHT RANGE OCCUPIED IN EACH PERIOD	Jan/Feb	Mar/Apr & Nov/Dec	May–Aug	Sep/Oct
	1.02–12 m	1.02–20 m	2.31–12 m	2.2–16 m

BAT SPECIES	ROOST PURPOSE			
	Winter	Transitory	Maternity	Mating
Barbastelle	✓	—	✓	—
Alcathoe's bat	—	—	✓	—
Daubenton's bat	—	✓	✓	—
Natterer's bat	—	—	—	✓
Leisler's bat	✓	✓	✓	✓
Noctule	✓	✓	—	✓
Nathusius' pipistrelle	—	✓	—	✓
Common pipistrelle	✓	✓	—	✓
Soprano pipistrelle	—	✓	—	—
Brown long-eared bat	—	✓	✓	✓

Hazard-beams: Top to bottom: noctules; common pipistrelle; and brown long-eared bats.

Table 3.14 Subsidence-cracks

CAUSED BY	The result of weight bearing down on a bent stem or limb (as opposed to the sudden and violent action of wind; hazard-beam, helical-split etc., or frost; frost-crack). The result is a split, much like a banana-skin when the banana is straightened			
DESCRIPTION	Subsidence-splitting appears more common on stems where the ground is seasonally damp, particularly where this situation has occurred more frequently as the tree has passed middle-age. On stems, subsidence results in a vertical split, often on opposing sides. Where the core rots-out they may comprise a tall and wide cylindrical void. On limbs, subsidence results in a down-turned trough; popular with dormice as well as bats. In stems, the roost position is typically in the apex. On limbs, the bats may be at the apex and typically nearest the stem, but not always			
TYPE	Crevice	✓		
	Void	✓		
APEX SHAPE	Dome	—		
	Peak/Wedge	✓		
	Spire	✓		
HABITAT	Continuous canopy	Woodland	✓	
		Plantation	✓	
		Orchard	—	
	Linear	Hedge / tree-line	—	
		Shelter-belt	—	
		Riparian fringe	—	
	Loose canopy	Parkland	✓	
		Churchyard & garden	—	
TOPOGRAPHY	Level	✓		
	Sloping	✓		
SITUATION	Central	✓		
	Corridor	✓		
	Edge	✓		
ROOST TREE SPECIES	Downy birch, beech, ash, sessile oak, pedunculate oak and crack willow			

DBH RANGE OF ROOST TREES OCCUPIED IN EACH PERIOD	Jan/Feb	Mar/Apr & Nov/Dec	May–Aug	Sep/Oct
	46 cm (data-deficient)	27.5–46 cm	15–56.4 cm	27 cm (data-deficient)

HEIGHT RANGE OCCUPIED IN EACH PERIOD	Jan/Feb	Mar/Apr & Nov/Dec	May–Aug	Sep/Oct
	2.46 m (data-deficient)	1.7-2.46 m	0.55–9 m	0.89 m (data-deficient)

BAT SPECIES	ROOST PURPOSE			
	Winter	Transitory	Maternity	Mating
Brandt's bat	Roosting identified but purpose unknown			
Natterer's bat	—	✓	✓	—
Noctule	—	—	✓*	—
Nathusius' pipistrelle	Roosting identified but purpose unknown			
Common pipistrelle	—	✓	—	—
Brown long-eared bat	—	✓	—	✓

*Squirrel-opened.

Subsidence-cracks: Top to bottom: noctules; brown long-eared bat; and brown long-eared bat and dormouse.

Table 3.15 Shearing-cracks

CAUSED BY	Wind twisting a limb or stem but not bending it			
DESCRIPTION	Very much a feature of limbs that are above 60°. The shearing-stress creates a vertically orientated crevice that extends into the core of the limb. These features are camouflaged as they are narrow (1–2 cm wide), and as the damage is linear within the natural stress-lines in the bark the woundwood seals the edges of the rend. Thus far the bat roost positions have been central along the line of the split and visible with a torch			
TYPE	**Crevice**	✓		
	Void	—		
APEX SHAPE	**Dome**	—		
	Peak/Wedge	✓		
	Spire	—		
HABITAT	**Continuous canopy**	Woodland		✓
		Plantation		—
		Orchard		—
	Linear	Hedge / tree-line		—
		Shelter-belt		—
		Riparian fringe		—
	Loose canopy	Parkland		—
		Churchyard & garden		—
TOPOGRAPHY	**Level**	✓		
	Sloping	✓		
SITUATION	**Central**	✓		
	Corridor	✓		
	Edge	✓		
ROOST TREE SPECIES	Pedunculate oak			

DBH RANGE OF ROOST TREES OCCUPIED IN EACH PERIOD	Jan/Feb	Mar/Apr & Nov/Dec	May–Aug	Sep/Oct
	—	71.3 cm (data-deficient)	71.3 cm (data-deficient)	100 cm (data-deficient)

HEIGHT RANGE OCCUPIED IN EACH PERIOD	Jan/Feb	Mar/Apr & Nov/Dec	May–Aug	Sep/Oct
	—	7.38 m (data-deficient)	7.38 m (data-deficient)	10 m (data-deficient)

BAT SPECIES	ROOST PURPOSE			
	Winter	Transitory	Maternity	Mating
Natterer's bat	—	✓	—	—
Leisler's bat	—	—	—	✓

Shearing-cracks: Top to bottom: Natterer's bats; and unknown

Table 3.16 Transverse-snaps

CAUSED BY	An impact, overloading or wind turbulence			
DESCRIPTION	Either a single significant split, or a ragged end with a series of crevices. Transverse-snaps can be on the stem or on a lateral limb and therefore either vertically or horizontally orientated, but regardless, the bat roost position is typically below or to one side of the entrance			
TYPE	Crevice	✓		
	Void	—		
APEX SHAPE	Dome	—		
	Peak/Wedge	✓		
	Spire	—		
HABITAT	Continuous canopy	Woodland	✓	
		Plantation	✓	
		Orchard	—	
	Linear	Hedge / tree-line	—	
		Shelter-belt	—	
		Riparian fringe	✓	
	Loose canopy	Parkland	—	
		Churchyard & garden	—	
TOPOGRAPHY	Level	✓		
	Sloping	✓		
SITUATION	Central	✓		
	Corridor	✓		
	Edge	✓		
ROOST TREE SPECIES	Beech, ash, sessile oak and pedunculate oak			

DBH RANGE OF ROOST TREES OCCUPIED IN EACH PERIOD	Jan/Feb	Mar/Apr & Nov/Dec	May–Aug	Sep/Oct
	118 cm (data-deficient)	62.8–118 cm	62.8–118 cm	35–100 cm

HEIGHT RANGE OCCUPIED IN EACH PERIOD	Jan/Feb	Mar/Apr & Nov/Dec	May–Aug	Sep/Oct
	8 m (data-deficient)	1.4–11 m	1.4–8 m	1.92–17 m

BAT SPECIES	ROOST PURPOSE			
	Winter	Transitory	Maternity	Mating
Barbastelle	—	✓	✓ (vertical)	—
Brandt's bat	—	—	✓ (vertical)	—
Leisler's bat	✓	✓	—	—
Common pipistrelle	✓	✓	—	✓
Soprano pipistrelle	✓	✓	—	—
Brown long-eared bat	—	✓	—	—

Transverse-snaps: Left: common pipistrelle; and right: brown long-eared bat.

Table 3.17 Welds

CAUSED BY	Two branches or stems crossing and fusing			
DESCRIPTION	Typically an almond-shaped pocket. Welds are vertically orientated but they may be accessed from the top or the bottom or even from the side. The roost position depends on the entrance			
TYPE	Crevice	✓		
	Void	✓ (large limbs only; both >20 cm diameter)		
APEX SHAPE	Dome	✓		
	Peak/Wedge	✓		
	Spire	—		
HABITAT	Continuous canopy	Woodland		✓
		Plantation		—
		Orchard		—
	Linear	Hedge / tree-line		✓
		Shelter-belt		—
		Riparian fringe		—
	Loose canopy	Parkland		✓
		Churchyard & garden		✓
TOPOGRAPHY	Level	✓		
	Sloping	✓		
SITUATION	Central	✓		
	Corridor	✓		
	Edge	✓		
ROOST TREE SPECIES	Beech, sessile oak, limes and Monterey cypress			

DBH RANGE OF ROOST TREES OCCUPIED IN EACH PERIOD	Jan/Feb	Mar/Apr & Nov/Dec	May–Aug	Sep/Oct
	28.5–170 cm	28.5–78.1 cm	28.5–104.4 cm	—

HEIGHT RANGE OCCUPIED IN EACH PERIOD	Jan/Feb	Mar/Apr & Nov/Dec	May–Aug	Sep/Oct
	1.42–5.3 m	1.42–6.73 m	1.42–8.35 m	—

BAT SPECIES	ROOST PURPOSE			
	Winter	Transitory	Maternity	Mating
Barbastelle	—	—	✓	—
Leisler's bat	—	✓	—	—
Noctule	—	—	✓	—
Brown long-eared bat	—	—	✓	—

Welds: Top to bottom: barbastelles (top three photos: J. Russ); noctules; and brown long-eared bats.

Table 3.18 Lifting-bark

CAUSED BY	Senescence and death of a limb and even the entire tree		
DESCRIPTION	A crevice between the bark and the heartwood, which can be anywhere and occupied both on the vertical and on the horizontal (like a hammock). For lifting-bark to provide a PRF, all it must offer is sufficient shade to offer shelter in a darkened position. The roost position may be literally anywhere; above, below, to one side and at any angle within that range		

TYPE	Crevice	✓	
	Void	—	

APEX SHAPE	Dome	—	
	Peak/Wedge	✓	
	Spire	—	

HABITAT	**Continuous canopy**	Woodland	✓
		Plantation	✓*
		Orchard	—
	Linear	Hedge / tree-line	—
		Shelter-belt	—
		Riparian fringe	✓
	Loose canopy	Parkland	✓
		Churchyard & garden	—

TOPOGRAPHY	Level	✓	
	Sloping	✓	

SITUATION	Central	✓	
	Corridor	✓	
	Edge	✓	

ROOST TREE SPECIES	Sweet chestnut, ash, sessile oak, pedunculate oak and giant sequoia			

DBH RANGE OF ROOST TREES OCCUPIED IN EACH PERIOD	Jan/Feb	Mar/Apr & Nov/Dec	May–Aug	Sep/Oct
	65–189 cm	9.5–189 cm	9.5–189 cm	60–118 cm

HEIGHT RANGE OCCUPIED IN EACH PERIOD	Jan/Feb	Mar/Apr & Nov/Dec	May–Aug	Sep/Oct
	2–4.94 m	1.37–2.69 m	1.04–10.42 m	1.49–6 m

BAT SPECIES	ROOST PURPOSE			
	Winter	Transitory	Maternity	Mating
Barbastelle	✓	✓	✓	✓
Bechstein's bat	—	✓	—	—
Alcathoe's bat	—	✓	—	—
Brandt's bat	—	✓	✓	—
Daubenton's bat	—	✓	—	—
Whiskered bat	—	✓	✓	—
Leisler's bat	✓	✓	—	✓
Noctule	✓	—	—	—
Nathusius' pipistrelle	✓	✓	—	—
Common pipistrelle	—	✓	—	—
Soprano pipistrelle	—	✓	—	✓
Brown long-eared bat	—	✓	—	—

*Including low-diameter weed species.

Lifting-bark: Top to bottom: barbastelle; Natterer's bats; and brown long-eared bat.

Table 3.19 Desiccation-fissures

CAUSED BY	Exposed desiccating and shrinking heartwood			
DESCRIPTION	Desiccation-fissures are radial fissures, typically narrow (1–2 cm wide), and may be deep, and only occur in dead wood. They are most common on heartwood tree species. Desiccation-fissures are in fact similar to shearing-cracks, but the way the features form is different and the two may be told apart by the significant area of exposed heartwood within which the desiccation is conspicuous, which is not present on shearing-cracks that have bark and woundwood bounding the latter crevice. Furthermore, shearing is more a feature of larger limbs, where desiccation-fissures are more common to the stem. Desiccation-fissures are always vertical; following the channels in the ray tissue. The roost position is typically (but not always) at the mid-point or above, and the bat(s) may be facing up or down			
TYPE	Crevice	✓		
	Void	—		
APEX SHAPE	Dome	—		
	Peak/Wedge	✓		
	Spire	—		
HABITAT	Continuous canopy	Woodland	✓	
		Plantation	—	
		Orchard	—	
	Linear	Hedge / tree-line	—	
		Shelter-belt	—	
		Riparian fringe	—	
	Loose canopy	Parkland	✓	
		Churchyard & garden	—	
TOPOGRAPHY	Level	✓		
	Sloping	—		
SITUATION	Central	✓		
	Corridor	✓		
	Edge	✓		
ROOST TREE SPECIES	Sweet chestnut, pedunculate oak and giant redwood			

DBH RANGE OF ROOST TREES OCCUPIED IN EACH PERIOD	Jan/Feb	Mar/Apr & Nov/Dec	May–Aug	Sep/Oct
	55.3–170 cm	30–170 cm	170 cm (data-deficient)	80 cm (data-deficient)*

HEIGHT RANGE OCCUPIED IN EACH PERIOD	Jan/Feb	Mar/Apr & Nov/Dec	May–Aug	Sep/Oct
	0.92–3.33 m	0.93–8.5 m	1.63 m (data-deficient)	13-16 m (data-deficient)*

BAT SPECIES	ROOST PURPOSE			
	Winter	Transitory	Maternity	Mating
Barbastelle	✓	✓	—	—
Leisler's bat	—	✓	—	—
Common pipistrelle	✓	—	—	—
Soprano pipistrelle	—	✓	—	—

*Although occupied during the mating-season, this was an individual transitory Leisler's bat.

Desiccation-fissures: Top to bottom: barbastelle; barbastelle and common pipistrelle; and common and soprano pipistrelles.

Table 3.20 Frost-cracks

CAUSED BY	Localised freeze-drying on one side of the stem alone. The result is a drying of the wood accompanied by contraction on one side, and expansion on the other. As only the outer layer of the tree is frozen, it is contracted against the inner core and thereby rent apart			
DESCRIPTION	Either a crevice with a peak or a cylindrical void with a spire apex. Although they can and do occur in many habitats, and may be a feature of heathland (particularly where the trees are also subject to drought in summer), frost-cracks are significantly more abundant on the north-eastern, eastern, south-eastern and southern sides of wooded ravines, where found on the uphill side. Frost-cracks are always in the stem and always vertical with a roost position that is either above the entrance, or to the side of the entrance behind a ram's-horn			

TYPE	Crevice	✓			
	Void	✓			

APEX SHAPE	Dome	—			
	Peak/Wedge	✓			
	Spire	✓			

HABITAT	Continuous canopy	Woodland	✓
		Plantation	—
		Orchard	—
	Linear	Hedge / tree-line	—
		Shelter-belt	—
		Riparian fringe	—
	Loose canopy	Parkland	—
		Churchyard & garden	—

TOPOGRAPHY	Level	✓ (rarely)	
	Sloping	✓	

SITUATION	Central	✓
	Corridor	—
	Edge	—

ROOST TREE SPECIES	Horse chestnut, downy birch, beech and sessile oak			

DBH RANGE OF ROOST TREES OCCUPIED IN EACH PERIOD	Jan/Feb	Mar/Apr & Nov/Dec	May–Aug	Sep/Oct
	8.6–57.7 cm	8.6–58 cm	8.6–58 cm	11–50.3 cm

HEIGHT RANGE OCCUPIED IN EACH PERIOD	Jan/Feb	Mar/Apr & Nov/Dec	May–Aug	Sep/Oct
	0.45–3.41 m	0.33–6.3 m	0.42–5.08 m	0.45–6.3 m

BAT SPECIES	ROOST PURPOSE			
	Winter	Transitory	Maternity	Mating
Barbastelle	—	✓	—	✓
Bechstein's bat	—	✓	—	—
Daubenton's bat	—	✓	—	—
Natterer's bat	✓	✓	—	✓
Leisler's bat	—	✓	—	—
Noctule	✓	✓	—	✓
Nathusius' pipistrelle	—	✓	—	—
Common pipistrelle	✓	✓	—	✓
Soprano pipistrelle	✓	✓	—	—
Brown long-eared bat	✓	✓	✓	✓

Frost-cracks: Top to bottom: barbastelle and brown long-eared bat; Leisler's bat and noctule; and Natterer's bat and common pipistrelles.

3.3.2 Summary

Table 3.21 illustrates which features comprise crevices, voids or both, and summarises the bat species that have been recorded exploiting the individual Damage PRFs. Importantly, the table also identifies the PRFs for which there are insufficient grounds to suggest that exploitation by an individual bat species might be 'more-likely-than-not'.

Table 3.21 An illustration of which Damage PRFs comprise crevices, voids or both, and which bat species have been recorded roosting in the individual PRF types and for what purpose

DAMAGE PRF SUMMARY	CREVICE	VOID	Barbastelle	Bechstein's bat	Alcathoe's bat	Brandt's bat	Daubenton's bat	Whiskered bat	Natterer's bat	Leisler's bat	Noctule	Nathusius' pipistrelle	Common pipistrelle	Soprano pipistrelle	Brown long-eared bat	Lesser horseshoe-bat
Lightning-strikes	✓	—	PN	—	—	—	PN	—	T	M	—	—	—	—	PN T	—
Hazard-beams	✓	✓	PN W	—	PN	—	PN T	—	M	PN T M W	T M W	T M	T M W	T	PN T M	—
Subsidence-cracks	✓	✓	—	—	—	?	—	—	PN T	—	PN	?	—	—	T M	—
Shearing-cracks	✓	—	—	—	—	—	—	—	T	M	—	—	—	—	—	—
Transverse-snaps	✓	—	PN T	—	—	PN	—	—	—	T W	—	—	T M W	T W	T	—
Welds	✓	✓	PN	—	—	—	—	—	—	T	PN	—	—	—	PN	—
Lifting-bark	✓	—	PN T M W	T	T	PN T	T	PN T	—	T M W	W	T W	T	T M	T	—
Desiccation-fissures	✓	—	T W	—	—	—	—	—	—	—	T	—	—	W	T	—
Frost-cracks	✓	✓	T M	T	—	—	T	—	T M W	—	T M W	T	T M W	T W	PN T M W	—

PN, occupation by a maternity colony within the Pregnancy and Nursery periods; T, occupation by 1–3 bats during the Pregnancy, Nursery and both Flux periods; M, occupation by any number of bats during the Mating period; W, occupation by any number of bats during the Winter period; ?, PRF is proven to be exploited by the bat species but it is unknown in which period usage was recorded.

The Damage PRFs comprise a significantly greater number of crevice-type PRFs.

Table 3.22 summarises the habitat types in which the individual Damage PRFs have been recorded and exploited by roosting bats.

Table 3.22 The habitat types in which the individual Damage PRFs have been recorded and exploited by roosting bats

CANOPY COVER	HABITAT TYPE	PRF TYPE								
		Lightning-strikes	Hazard-beams	Subsidence-cracks	Shearing-cracks	Transverse-snaps	Welds	Lifting-bark	Desiccation-fissures	Frost-cracks
CONTINUOUS CANOPY	Woodland	✓	✓	✓	✓	✓	✓	✓	✓	✓
	Plantation	—	—	✓	—	✓	—	✓	—	—
	Orchard	—	—	—	—	—	—	—	—	—
LINEAR CANOPY	Hedge/tree-line	—	✓	—	—	—	✓	—	—	—
	Shelter-belt	—	—	—	—	—	—	—	—	—
	Riparian-fringe	—	✓	—	—	—	—	—	—	—
LOOSE CANOPY	Parkland	✓	✓	✓	—	—	✓	✓	✓	—
	Churchyard & garden	—	✓	—	—	—	✓	—	—	—

3.4 The Association PRFs

The Association PRFs are an interesting pair: one is an often overlooked but very profitable PRF, the other is contentious in the extreme.

Accounts of each of the Association PRFs are provided in Table 3.23 to Table 3.26.

Table 3.23 Fluting

CAUSED BY	A natural and entirely normal formation		
DESCRIPTION	Flutes are longitudinal grooves or channels in the stem. The PRF offered is typically a pocket within the tissue. Although they may extend up or down, where they simply go in for sufficient distance to offer a dark cleft that is sheltered from rain and wind, this appears sufficient for several bat species. Vertically orientated and all records thus far are from within the field-layer. The bat roost position is typically to the side of the entrance and clearly visible with a torch alone		

TYPE	Crevice	✓	
	Void	—	
APEX SHAPE	Dome	—	
	Peak/Wedge	✓	
	Spire	—	

HABITAT	Continuous canopy	Woodland	✓
		Plantation	—
		Orchard	—
	Linear	Hedge / tree-line	—
		Shelter-belt	—
		Riparian fringe	—
	Loose canopy	Parkland	✓
		Churchyard & garden	✓ (churchyard)

TOPOGRAPHY	Level	✓	
	Sloping	—	
SITUATION	Central	✓	
	Corridor	✓	
	Edge	—	

ROOST TREE SPECIES	Hornbeam, beech, black locust and yew			
DBH RANGE OF ROOST TREES OCCUPIED IN EACH PERIOD	Jan/Feb	Mar/Apr & Nov/Dec	May–Aug	Sep/Oct
	43–112.5 cm	43–82 cm	43–55 cm	49–82 cm
HEIGHT RANGE OCCUPIED IN EACH PERIOD	Jan/Feb	Mar/Apr & Nov/Dec	May–Aug	Sep/Oct
	0.58–2.2 m	0.58–1.9 m	0.82–1.75 m	0.98–1.9 m

BAT SPECIES	ROOST PURPOSE			
	Winter	Transitory	Maternity	Mating
Barbastelle	—	✓	—	—
Natterer's bat	—	✓	—	✓
Common pipistrelle	✓	✓	—	—
Soprano pipistrelle	—	✓	—	✓
Brown long-eared bat	✓	✓	—	✓

Fluting: Top to bottom: Natterer's bats; brown long-eared bat (middle three photos: R. Koczy); and brown long-eared bat.

Table 3.24 Ivy

CAUSED BY	Ivy is a climbing (and occasionally carpeting) plant that that can grow to 30 m and is common in woodlands, scrub and hedgerow trees on neutral soils
DESCRIPTION	A nebulous web of crevices with its own evergreen screen and sheltered microclimate that encompasses every orientation and aspect from the ground-layer and into the canopy. A widespread plant, it is listed in all but three of the 19 NVC woodland communities (the exceptions being W11, 18 and 19). The use of ivy by roosting bats is still imperfectly understood and a useful description is therefore impossible. The photographic evidence comprises two encounters. The first was captured using thermal imaging by BSG Ecology and relates to three common pipistrelles that were sheltering in a dense table of dead low-diameter ivy stems in a sunlit position on a woodland tree (G. Miller, personal communication, April 2016). The second was captured using an ordinary digital camera and relates to an individual barbastelle hiding in plain sight behind a large-diameter ivy stem on a sunlit oak. Despite the obvious differences in the ivy in the two situations, there is one similarity; it was open to the sun

TYPE	Crevice	✓		
	Void	—		

APEX SHAPE	Dome	—		
	Peak/Wedge	✓		
	Spire	—		

HABITAT	**Continuous canopy**	Woodland	✓
		Plantation	—
		Orchard	—
	Linear	Hedge / tree-line	—
		Shelter-belt	—
		Riparian fringe	—
	Loose canopy	Parkland	—
		Churchyard & garden	—

TOPOGRAPHY	Level	✓
	Sloping	—

SITUATION	Central	—
	Corridor	—
	Edge	✓

ROOST TREE SPECIES	Pedunculate oak and limes			

DBH RANGE OF ROOST TREES OCCUPIED IN EACH PERIOD	**Jan/Feb**	**Mar/Apr & Nov/Dec**	**May–Aug**	**Sep/Oct**
	—	—	78.1–148.6 cm	78.1–148.6 cm

HEIGHT RANGE OCCUPIED IN EACH PERIOD	**Jan/Feb**	**Mar/Apr & Nov/Dec**	**May–Aug**	**Sep/Oct**
	—	—	1.12–1.49 m	1.12–1.49 m

BAT SPECIES	**ROOST PURPOSE**			
	Winter	**Transitory**	**Maternity**	**Mating**
Barbastelle	—	✓	—	—
Common pipistrelle	—	✓	—	—

Ivy: Top: barbastelle; and bottom: common pipistrelles (bottom two photos: BSG Ecology – found using a FLIR 650 thermal Imaging camera).

3.4.1 Summary

Table 3.25 illustrates which features comprise crevices, voids or both, and summarises the bat species that have been recorded exploiting the individual Association PRF. Importantly, the table also identifies the PRF for which there are insufficient grounds to suggest that exploitation by an individual bat species might be 'more-likely-than-not'.

Table 3.25 An illustration of which Association PRFs comprise crevices, voids or both, and which bat species have been recorded roosting in the individual PRF types and for what purpose

ASSOCIATION PRF SUMMARY	CREVICE	VOID	Barbastelle	Bechstein's bat	Alcathoe's bat	Brandt's bat	Daubenton's bat	Whiskered bat	Natterer's bat	Leisler's bat	Noctule	Nathusius' pipistrelle	Common pipistrelle	Soprano pipistrelle	Brown long-eared bat	Lesser horseshoe-bat
Fluting	✓	—	T	—	—	—	—	—	T M	—	—	—	T	T M	T M W	—
Ivy	✓	—	T	—	—	—	—	—	—	—	—	—	—	T	—	—

T, occupation by 1–3 bats during the Pregnancy, Nursery and both Flux periods; M, occupation by any number of bats during the Mating period; W, occupation by any number of bats during the Winter period.

The Association PRFs comprise crevice-type PRFs alone.

Table 3.26 summarises the habitat types in which the individual Association PRFs have been recorded and exploited by roosting bats.

Table 3.26 The habitat types in which the individual Association PRF have been recorded and exploited by roosting bats

CANOPY COVER	HABITAT TYPE	PRF TYPE	
		Flutes	Ivy
CONTINUOUS CANOPY	Woodland	✓	✓
	Plantation	—	—
	Orchard	—	—
LINEAR CANOPY	Hedge/tree-line	—	—
	Shelter-belt	—	—
	Riparian-fringe	—	—
LOOSE CANOPY	Parkland	✓	—
	Churchyard & garden	✓	—

Intelligence-Gathering

4.1 Introduction

The objective of intelligence-gathering is to collect and collate the evidence required to perform a reasoned appraisal of whether there are grounds to suggest it is 'more-likely-than-not' a site may hold Potential Roost Features (PRFs) that might be exploited by roosting bats.

From a consultancy perspective, National Planning Practice Guidance identifies that *'an ecological survey will be necessary if the type and location of development are such that the impact on biodiversity may be significant and existing information is lacking or inadequate'*.

Intelligence-gathering is simply the collection and collation of the evidence required to prove whether or not there are grounds to suggest action will be 'more-likely-than-not' to be worthwhile. It should be noted that the interpretation of results is a separate stage, and will be discussed in the next chapter.

Regardless of whether the search for tree-roosts is being performed in support of a planning application, to inform a management plan, or by an amateur naturalist, intelligence-gathering is where every search should begin.

Being disciplined and gathering the intelligence in advance of every search for tree-roosts is what separates those who will not only find tree-roosts, but will learn and progress to a level where they command respect, from those others who, in the words of Sir Walter Scott, 'shall go down … unwept, unhonored, and unsung'.

Intelligence-gathering is not solely a desk-based exercise, but requires reconnaissance for it to be of practical use. However, if there is a motto for intelligence-gathering it is this:

Simplicity; simplicity; and simplicity.[1]

---

1 Henry Thoreau suggested that by adopting this maxim, the individual is able to more fully connect with the universe around them and learn the secrets of existence.

4.2 Resources

In a professional context, the first stage of intelligence-gathering is to establish whether there is any possibility of a 'show-stopping' situation.

It is easy to overlook the fact that intelligence-gathering has two directions, and establishing the client's position is every bit as important as establishing the location of the bats. The presence of a colony of a particular bat species may be such as to make the development proposed impossible in terms of compensation, impractical in terms of cost, or undesirable in terms of potentially bad publicity. Where the potential an insurmountable obstacle exists can be reasonably predicted, explaining the gravity of the situation to the client at the outset allows them to assess whether or not the investment is worth the risk, and ensures the development design team are thinking of the potential mitigation and compensation that might be required.

The second stage of intelligence-gathering is to establish whether sufficient information already exists to provide a robust account of the likely tree-roost interest within the site.

Referring to the need for a proportionate approach identified in Chapter 1, the detail and depth of the information required will depend on what motivates the search. However, to achieve a reasoned assessment of the likelihood an area of wooded habitat will hold PRFs, four resources are always useful, comprising:

1. A map of the location and extent of wooded habitat and its situation in the wider landscape.
2. A description of the wooded habitat species composition and structure.
3. An inventory of bat species present within the county.
4. The results of a data-search for historic bat data.

4.2.1 The habitat map

The habitat map identifies the location and extent of the wooded habitat that is to be searched, the connectivity of that habitat to hunting grounds and to other areas of roost habitat, and the topographical situation.

Different bat species occur in different wooded habitats and have different sensitivities to roost isolation within a bare landscape. Topography influences the different PRFs that may occur. The habitat map may therefore be used to identify whether there are grounds to suggest bat species might exploit the site. In addition, the habitat map will assist the survey team in the decision of how to map the individual host trees.

A habitat map is best created by using all the following resources:

» The interactive maps provided at http://www.magic.gov.uk/.
» Phase 1 habitat mapping (JNCC 2010),[2] (typically performed at the outset of a Preliminary Ecological Appraisal).
» Satellite imagery (such as Google Earth, and www.gridreferencefinder.com).
» Ordnance Survey mapping.

Where different types of habitat exist, it is helpful to classify them into the broad habitat types identified at Chapter 2.

Magic.gov.uk

The MAGIC website is managed by Natural England (the government's adviser for the natural environment in England) and provides authoritative geographic information

---

2 Phase 1 habitat mapping was devised by the Joint Nature Conservation Committee (JNCC) as a standardised system for classifying and mapping wildlife habitats (JNCC 2010).

about the natural environment. The information includes all wooded habitats across Great Britain and is presented in an interactive map which can be explored using various mapping tools that are included.

Phase 1 habitat mapping
Phase 1 habitat mapping separates woodland into broadleaved and coniferous, and semi-natural from plantation. It also identifies parkland and scrub, and hedges with trees from those without. Individual larger trees may also be identified as 'Target Notes' (although there is great variation between surveyors; typically, the smaller the site, the more diligently the trees are noted).

Satellite imagery
Satellite imagery is useful in that it illustrates the context in which the wooded habitat exists within the wider countryside, and the connectivity of individual parcels of wooded habitat. Satellite imagery can also be used to assess canopy cover and relative structure between individual habitat components, as well as identifying how many trees are in a park or on a hedgerow.

Ordnance Survey mapping
Ordnance Survey mapping divides woodland into coniferous and broadleaved, and identifies orchards, but that is all. The greatest value of Ordnance Survey mapping is that it gives an accurate representation of the topography.

4.2.2 The habitat description
Different habitats hold different tree species which form different PRF types at different ages, and are exploited by different bat species at different times of year for different purposes. In broad terms, tree species composition and density influence the abundance of Disease & Decay and Association PRFs, topography and structure influence the abundance of Damage PRFs.

It may be that a description already exists of the wooded habitat(s) that are present in the site within a Preliminary Ecological Appraisal (PEA)[3] or, if the site is a Statutory or non-Statutory Wildlife Site, in a notification/citation. It may be that the site is listed on the Ancient Woodland Inventory as Ancient Semi-Natural Woodland or another variation. However, unless the PEA was written by another naturalist with an interest in tree-roosting ecology, it is unlikely the description will be of any practical use. All the other descriptive accounts are 'static' (i.e. background noise).

For a description to be of practical use in a search for tree-roosts, it will comprise the following:

» **Topography** – whether the wooded habitat comprises:
   a. Predominantly level ground; and
   b. Sloping ground.
» **Situation** – whether the wooded habitat comprises:
   a. Trees in *central* situations (typically woodland and plantation);
   b. Trees in *corridor* situations (i.e. on the banks of a river, sides of a bridleway/footpath/deer-path through a wood); and
   c. Trees in *edge* situations (i.e. with open ground on at least one side).

---

3   An assessment undertaken to inform a Planning Application. Although these assessments include broad habitat descriptions and botanical species lists, unfortunately it is rare for the description to include structural information such as the canopy height of woodland or individual trees, or even an approximate range of the diameter at breast height within a stand.

» **Composition** – which tree species are present in what abundance and what size, as follows:

 a. Canopy-layer:

  – A list of the canopy-layer tree species present;

  – An estimate of the relative abundance of each species;

  – An estimate of the average height of the overall canopy to the nearest metre; and

  – The average diameter at breast height (DBH) of each species present to the nearest centimetre.

 b. Shrub-layer:

  – A list of the shrub-layer[4] species present;

  – An estimate of the relative abundance of each species; and

  – The average DBH of each species present to the nearest centimetre.

In all probability, this will require specific reconnaissance, but this need not be onerous; the habitat description is not a detailed botanical description and is most usefully set out in tabulation form (as will be illustrated at the close of the chapter). Reconnaissance is doubly helpful in that the habitat map can be 'truthed' for accuracy at the same time.

Topography

The broad topography of an area of habitat can be readily deduced from the gradient lines on Ordnance Survey sheets. However, when on the ground, it is useful to bear in mind that the gradient influences the PRF forms that may occur due to the movement of air in response to night-time temperature, and exposure to wind. If the trees are growing on level ground, then night-time air movement is minimal, those at the edge are exposed to extremes of wind but those in the middle are sheltered. If the trees are on sloping ground, then night-time air movement will be more acute because hot air in the lower levels will rise and this will draw down cooler air on the upper levels. In addition, the more acute the slope, the greater the canopy 'edge' will be and therefore the more exposed to wind.

Figure 4.1 Degrees of slope

All that is required is a very basic characterisation. Looking at Figure 4.1, 0–10° is broadly level. Anything above 20° is significantly sloping and would begin to result in noticeable night-time air movement with colder air flowing down the slope displacing warmer air at the base. Anything above 40° is sufficiently acute that all the trees (i.e. not only those on the woodland margin) would be susceptible to wind-damage in a storm.

---

4 This does not mean that only the 'shrub' species are recorded; all the woody species that make up the layer should be recorded. It is not uncommon for the same species that are present in the canopy to make up a significant proportion of the shrub-layer too.

Abundance

Abundance is best defined using the DAFOR scale, which classifies species as follows:

» Dominant.

» Abundant.

» Frequent.

» Occasional.

» Rare.

The DAFOR scale can be used to estimate the density of each tree species, or the proportion of the canopy they occupy. The trick here is not to get too bogged down in percentages; it's subjective anyway but it really doesn't have to be accurate to the nth degree for the analysis to be meaningful.

A site may not hold species in all abundance classes. When using this measure what typically happens is that you either get one or other of these three situations:

» A dominant canopy species potentially with additional occasional and rare species (think coppice with standards, plantation, orchard, shelter-belt and hedgerow).

» Two abundant species with occasional and rare species (typical semi-natural woodland).

» Three or more frequent species with occasional and rare species (riparian-fringe, parkland).

If you have only got one tree, and it is quite clearly an isolated individual, then frankly this aspect is irrelevant. But if you are working on one tree that is visibly amongst several more, then consider what the context is and classify accordingly.

The reason for recording all this is to be consciously aware of what it is you are seeing. What you are doing here is programming your episodic memory.

At this stage, you are not looking for PRFs and it is better that you keep your focus at a wide-angle and just take in the context. As a result, you can do this reconnaissance at any time of year. Obviously if you find a PRF it would be wise to note it, but finding PRF is not the objective of the exercise.

An incidental advantage is that several sites can be scoped in a single day to select the one that is most likely to be profitable for an amateur study.

Canopy height

A clinometer is used to measure the canopy height. However, testing has shown that the canopy height does not need to be accurate to the nth degree; an average will be adequate. In addition, the shrub-layer height is not needed for the analysis at the desk-study stage, all that is required is the vertical range of the overall habitat structure – lowest to highest.

Diameter at breast height

The DBH is recorded with a diameter tape or tree calipers. Two figures are needed to provide a robust representation of the minimum and maximum DBH range of each tree species present. This does not mean that every tree should be measured, but that a representative sample that illustrates the 'ball-park' for both ends of the scale should be chosen – lowest to highest diameter.

4.2.3 The bat species inventory

The bat species inventory is simply a list of the bat species that occur in the county. This can be gained from the Local 'Biological' Records Centre (LRC) or the local Bat Group (which may have an inventory list on their website). The list may then be narrowed down to comprise only those that roost in trees. It is helpful to order the list from the commonest species to the rarest, as with Table 1.1 in Chapter 1.

Different suites of bat species occur in different counties. The bat species inventory is therefore particularly helpful when the search is to be performed in a locality with which the survey team will not be familiar. Reading the list carefully and identifying the bat species that are less familiar to the survey team, in order that their roosting ecology can be reviewed in detail prior to commencing any subsequent surveillance campaign, consistently rewards in proportion to the revision effort. Even the act of producing the list is beneficial in terms of galvanising focus and should not be underestimated.

4.2.4 The data-search

Most counties have a LRC and the data-search is a search of the county biological records.

A biological record is a pithy account of an encounter with a floral or faunal species and comprises:

» The species.

» The date.

» The location.

» The recorder[5].

If you are an amateur, bat data can usually be obtained from the LRC free of charge. If you are a professional, you must pay for the time it takes the clerk to perform the search. As off-site in-flight data takes an aeon to wade through and is typically a mass of individual incomplete records that have little or no merit, when a data-search is to be used to inform a search for tree-roosts, it is most helpful when it is stratified as follows:

» On-site roost and in-flight records.

» Off-site roost records alone that fall within the species average Core Sustenance Zones defined by Bat Conservation Trust (Collins 2016), which are as follows:

 – Bechstein's bat *Myotis bechsteinii*, Alcathoe's bat[6] *M. alcathoe*, Brandt's bat *M. brandtii* and whiskered bat *M. mystacinus* – **within a 1 km radius;**

 – Daubenton's bat *Myotis daubentonii*, common pipistrelle *Pipistrellus pipistrellus* and lesser horseshoe-bat *Rhinolophus hipposideros* – **within a 2 km radius;**

 – Leisler's bat *Nyctalus leisleri*, Nathusius' pipistrelle *Pipistrellus nathusii*, soprano pipistrelle *P. pygmaeus* and brown long-eared bat *Plecotus auritus* – **within a 3 km radius;**

 – Natterer's bat *Myotis nattereri* and noctule *Nyctalus noctula* – **within a 4 km radius;** and

 – Barbastelle *Barbastella barbastellus* – **within a 6 km radius.**

---

5   Do not expect to get the recorder's name with the record. Why this is withheld we neither know nor care, it just is. However, if you are seeking the background to a sequence of roost-records at a specific location, this can usually be obtained from the Secretary of the Local Bat Group, most of whom can be contacted online though the group's website. In this situation, it is courteous to offer your data in return.

6   Alcathoe's bat has no Core Sustenance Zone defined and is therefore arbitrarily assigned that defined for Bechstein's bat, Brandt's bat and whiskered bat.

Note: if you are working within the distribution zones of the barbastelle, Bechstein's bat or the lesser horseshoe-bat, it is wise to ask the LRC for any Sites of Special Scientific Interest (SSSI)[7] and Special Area of Conservation (SAC)[8] notifications in respect of bats that fall within the data-search radii for those species. If any such notifications exist, search the Natural England Research Reports online using the site name and the bat species and you may find reports of radiotracking studies done in respect of those notifications. As these reports contain maps showing where the bats roosted and foraged, they will inform you if any of the bats tracked entered your site.

In some cases, the exact location of roost-records may be withheld and the LRC may simply identify that a roost of the species exists within the search radii. This is not a problem; all you need to know is whether there is historic evidence the species has roosted within your site, and whether your site is within the Core Sustenance Zone of a known off-site roost; you are ticking a box here, that is all.

4.3 Collation of the results

The result of the intelligence-gathering should be a map, and an accompanying table, something along the lines of Table 4.1 (a Word copy of which can be downloaded from www.battreehabitatkey.com).

This is all you need to complete the desk-study and decide whether there are grounds to suggest it might be 'more-likely-than-not' that the site will hold PRF and which bats are 'more-likely-than-not' to exploit those PRF as roosts.

---

7   Sites of Special Scientific Interest (SSSIs) are sites that are legally protected for the habitats and/ or species they hold.

8   Special Areas of Conservation (SACs) comprise an individual SSSI, or a suite of SSSIs that in combination receive additional protection under the Habitats Directive (one of the EU's two directives in relation to wildlife and nature conservation). Three tree-roosting bat species are listed on Annex II of the Habitats Directive: the barbastelle, Bechstein's bat and lesser horseshoe-bat. These three species have SACs that are notified to protect key elements of their habitat. In addition, some SACs list one or more Annex II bat species as a 'qualifying feature' even when they are not the primary reason for the notification.

Table 4.1 The results of intelligence-gathering

HABITAT TYPE		
TOPOGRAPHY	Level (i.e. 0–10°)	
	Sloping (i.e. over 10°)	
SITUATION	Central	
	Corridor	
	Edge	
CANOPY HEIGHT RANGE		
CANOPY TREE SPECIES	ABUNDANCE	DBH RANGE
SHRUB-LAYER TREE SPECIES	ABUNDANCE	DBH RANGE
BAT SPECIES INVENTORY (DATA-SEARCH)		

Bat species columns:
- Lesser horseshoe-bat
- Brown long-eared bat
- Soprano pipistrelle
- Common pipistrelle
- Nathusius' pipistrelle
- Noctule
- Leisler's bat
- Natterer's bat
- Whiskered bat
- Daubenton's bat
- Brandt's bat
- Alcathoe's bat
- Bechstein's bat
- Barbastelle

Bat species inventory rows:
- BAT SPECIES PRESENT IN COUNTY
- BAT SPECIES RECORDED ROOSTING ON-SITE
- BAT SPECIES RECORDED IN-FLIGHT ON-SITE
- BAT SPECIES RECORDED ROOSTING OFF-SITE AND FOR WHICH THE SITE IS WITHIN THE CORE SUSTENANCE ZONE

CHAPTER 5

The Desk-Study

In this chapter	
Introduction	The desk-study rationale
The desk-study objective	The questions the desk-study framework is designed to answer
The test of *"reasonable likelihood"*	The interpretation of the desk-study against the threshold of 'more-likely-than-not', and the collation of the results into a format that has a practical application
To survey or not to survey?	Consideration of the Natural England's European Protected Species Licensing (EPSL) Policy 4 criteria (Natural England 2016)

5.1 Introduction

The desk-study is the review and interpretation of the data collated at the intelligence-gathering stage within a repeatable framework, in order to answer the question of whether further survey action would be 'more-likely-than-not' to be rewarded with a positive outcome, i.e. the finding of a tree-roost.

When the search is being performed for research or simply for pleasure, the rationale behind the desk-study is that it enables the filtering of sites to target those that will most likely be profitable.

When the search is being performed professionally, as what might be thought of as a 'biological risk assessment', the rationale is that it provides an evidence-base in support of recommendations that will, on one hand guard, against potentially unreasonable demands for further action, whilst also guarding against any suggestion of sharp practice,[1] and on the other, will support recommendations for action with a detailed hypothesis encompassing:

» The tree species that are 'more-likely-than-not' to hold Potential Roost Features (PRFs).

---

1   To illustrate, if the desk-study concludes that there is insufficient evidence to support a hypothesis that the site has a *"reasonable likelihood"* of holding PRFs, and that even if PRFs do in fact exist, there is insufficient evidence to suggest a *"reasonable likelihood"* that species of bats that are known to exploit those PRFs will encounter them, then the motive behind any recommendation for a survey by a consultancy that will result in a financial gain to that consultancy might reasonably be questioned.

» The diameter of the individual specimens of those trees that are 'more-likely-than-not' to hold PRFs.

» The PRF forms that are 'more-likely-than-not' to be present.

» The bat species that are 'more-likely-than-not' to exploit the PRF (as well as those that might also potentially be encountered).

However, in a professional situation, even where the results of the desk-study indicate that further action would be warranted, this is not a springboard into one single operation but a progression of operations that may conclude at more than one point. This is because, even if a greater depth of resolution is considered proportionate, there are situations where the campaign may be fulfilled without the need for surveillance; by far the greater proportion of professional appraisals are in support of an Ecological Impact Assessment, and the progression of operations will encompass the following three stages:

1. The truthing of the desk-study conclusions, by making a detailed ground inspection of the trees to establish whether they really do hold PRFs.

   and if they do

2. An initial inspection of any PRFs encountered in order to establish whether they are actually suitable as roost sites.

   and if they are

3. The design and performance of an effective surveillance programme to establish whether those PRF are exploited as roost sites.

If the answer to either stage 1 or stage 2 is no, then obviously no surveillance will be required.

5.2 The desk-study objective

The objective of the desk-study is to answer the following three questions:

» **Question 1:** Are there grounds to suggest that it is 'more-likely-than-not' that PRFs will be present and of a character known to be exploited by bat species that are known to have colonies within range of the site, and might therefore be 'more-likely-than-not' to encounter those PRFs if they are present?

   And, if the threshold is crossed

» **Question 2:** Is there sufficient information to satisfy Natural England's EPSL Policy 4 criteria (Natural England 2016), and thereby decide whether further action would be proportionate to the level of risk?

   And, if there is not

» **Question 3:** *Specifically*, what additional information is required?

Ultimately, at the end of the desk-study there should be a framework of information that demonstrates unequivocally that further action is, or is not, required. If further action is considered justified, the desk-study should provide a summary of the results in a format that can be used to inform the next stage by defining the objective 'question(s)' (i.e. the hypothesis) against which the results can be compared.

5.3 The test of *"reasonable likelihood"*

5.3.1 Rationale

In order to put the threshold of 'more-likely-than-not' back into the context of ODPM Circular 06/2005 and the National Planning Practice Guidelines criteria,[2] a test of *"reasonable likelihood"* may be performed using the broad environmental and habitat data collected at the intelligence-gathering stage. This might follow the four stages set out below and using a simple tabulation, an example of which is provided at Table 5.1 (a Word copy of this table can be downloaded from www.battreehabitatkey.com).

Stage 1

Using the tree species summary table (Table 5.1) create a summary for each tree species present in the site.

Firstly, using the *PRF Summary Tables* provided in Appendix A enter the minimum DBH the tree species has been proven to form each of the three PRF types: Disease & Decay, Damage, and Association.

Secondly, compare the DBH range present within the site against the minimum DBH the tree species forms PRF.

» If both aspects are satisfied and the species is present above the minimum DBH threshold, proceed to Stage 2.

» If the trees present are not tree species for which there is accessible evidence that a specimen has held roosting bats, or there are no specimens above the minimum DBH for which roost records exist on the BTHK Database, the justification for proceeding might reasonably be questioned.

Stage 2

Compare the results of the DBH range assessment against the *PRF Summary Tables* in Appendix A, to identify which PRF types there is available evidence that roosting bats have exploited and complete Stage 2 on the table. For example, if field maple is present above the minimum DBH threshold, then the boxes for knot-holes, tear-outs and wounds would be ticked.

Note: Some tree species are 'data deficient' in the *PRF Summary Tables*. To overcome this, the minimum DBH for that particular PRF type identified in the PRF accounts in Chapter 3 may be adopted. For example, elm *Ulmus* sp. is data deficient, but is known from photographic accounts to form weld PRFs that are exploited by roosting bats. As the minimum DBH that wounds are known to be exploited is 8.7 cm, this DBH threshold may be adopted in order to complete the test, the rationale being that this is currently the best evidence available.

Stage 3

The *PRF Summary Tables* identify all the PRF forms, but are an aggregate of: all habitat types; both topographical contexts; and, all three situations.

Therefore, further truthing will be required to 'scope-out' any PRFs that do not occur in the habitat, topography and situation the site encompasses. This can be performed by comparing each PRF type identified, and the habitat, topography and situation in which the tree species occurs, with the individual PRF descriptions provided in Chapter 3. If the PRF type does not occur within the environment present, it is scoped out. All those PRFs that remain are listed in the column headed 'PRF Potentially Present Within Site'.

---

2 See Chapter 1.

Table 5.1 Template for a tree species summary table

	TREE SPECIES				
STAGE 1	DIAMETER at BREAST HEIGHT				
	DBH RANGE OF SPECIMENS OF THE SPECIES ON SITE				
	Minimum DBH at which species is proven to form Disease & Decay PRF		Trees within DBH range present on site (Y/N?		
	Minimum DBH at which species is proven to form Damage PRF		Trees within DBH range present on site (Y/N)?		
	Minimum DBH at which species is proven to hold Association PRF		Trees within DBH range present on site (Y/N)?		

	PRF POTENTIALLY PRESENT THAT ROOSTING BATS HAVE BEEN PROVEN TO EXPLOIT					
STAGE 2	Disease & Decay	Y/N	Damage	Y/N	Association	Y/N
	Woodpecker-hole		Lightning-strike		Fluting	
	Squirrel-hole		Hazard-beam		Ivy	
	Knot-hole		Subsidence-crack			
	Pruning-cut		Shearing-crack			
	Tear-out		Transverse-snap			
	Wound		Weld			
	Canker		Lifting-bark			
	Compression-fork		Desiccation-fissure			
	Butt-rot		Frost-crack			

	ENVIRONMENT			
STAGE 3	HABITAT		Y/N	PRF POTENTIALLY PRESENT WITHIN SITE (list)
	TOPOGRAPHY	Level		
		Sloping		
	SITUATION	Central		
		Corridor		
		Edge		

	BAT SPECIES					
STAGE 4	Species proven to exploit PRF potentially present in this context	Y/N	County inventory	Y/N	Data-search result	Y/N
	Barbastelle		Barbastelle		Barbastelle	
	Bechstein's bat		Bechstein's bat		Bechstein's bat	
	Alcathoe's bat		Alcathoe's bat		Alcathoe's bat	
	Brandt's bat		Brandt's bat		Brandt's bat	
	Daubenton's bat		Daubenton's bat		Daubenton's bat	
	Whiskered bat		Whiskered bat		Whiskered bat	
	Natterer's bat		Natterer's bat		Natterer's bat	
	Leisler's bat		Leisler's bat		Leisler's bat	
	Noctule		Noctule		Noctule	
	Nathusius' pipistrelle		Nathusius' pipistrelle		Nathusius' pipistrelle	
	Common pipistrelle		Common pipistrelle		Common pipistrelle	
	Soprano pipistrelle		Soprano pipistrelle		Soprano pipistrelle	
	Brown long-eared bat		Brown long-eared bat		Brown long-eared bat	
	Lesser horseshoe-bat		Lesser horseshoe-bat		Lesser horseshoe-bat	

> » If specimens of the tree species occurring in the habitat, topography and situation encompassed by the site and tree species have been proven to have held specific PRF forms that were exploited by roosting bats, proceed to Stage 4.

> » If the combination of habitat, topography and situation present do not match those of roost records of that particular tree species held on the BTHK Database, the justification for proceeding might reasonably be questioned.

Stage 4

Firstly, compare the list of 'PRFs Potentially Present Within Site' with the PRF descriptions provided in Chapter 3 to identify which bat species have been proven to exploit the PRF types potentially present.

Secondly, using the 'Bat Species Inventory' collated at Table 4.1 at the intelligence-gathering stage, identify which of the species proven to exploit the PRFs potentially present occur in the county.

Thirdly, using the 'Bat Species Inventory' identify which bat species have been recorded within the site or within range of the site.

> » Those species that are present in the county that are proven to exploit the situation, environment and PRFs the site encompasses are species that might *potentially* exploit on-site PRFs to roost.

> » Those species that have colonies within range of the site may be concluded to be 'more-likely-than-not' to exploit suitable on-site PRFs for roosting.

> » If it is 'more-likely-than-not', *"reasonable likelihood"* has been demonstrated; proceed to Stage 5.

> » If it is not 'more-likely-than-not', *"reasonable likelihood"* has not been demonstrated; question the justification for proceeding.

Stage 5

Although the threshold of 'more-likely-than-not' is the threshold for further action, where that threshold has been crossed it does not mean that all other potential should be ignored.

The Tree Species Summary tables (as illustrated at Table 5.1) do not have a practical application. A Desk-study field summary is therefore required for use in the field and Table 5.2 is a template for this (a copy of this table can be downloaded from www.battreehabitatkey.com).

This is completed by cross-referencing Stages 3 and 4 in Table 5.2 with the PRF summaries provided in Chapter 3. Bat species that are 'more-likely-than-not' to be present are identified by the use of the suffix 'RL' (for *"reasonable likelihood"*), and bat species with a 'potential' likelihood of presence are identified by the use of the suffix 'P'.

The value of this exercise is that where ground-truthing is performed, the surveyor will be searching the individual tree species for the specific PRFs it is known to form, and that are known to be exploited by roosting bats in that situation.

The surveyor may therefore prepare in advance of the survey by familiarising them-selves with that PRF type; what it looks like, and where in the structure of the tree it is likely to occur. By so doing, they will be focusing on a limited number of PRF types, rather than trying to record everything that might occur. This does not mean that PRFs in atypical situations should automatically be scoped-out, but if the operation is being performed in a professional capacity, the principle of 'more-likely-than-not' should be applied at every stage.

Table 5.2 Desk-study field summary

	Disease & Decay									Damage									Association	
SITE NAME																				
TREE SPECIES																				
BAT SPECIES	Woodpecker-hole	Squirrel-hole	Knot-hole	Pruning-cut	Tear-out	Wound	Canker	Compression-fork	Butt-rot	Lightning-strike	Hazard-beam	Subsidence-crack	Shearing-crack	Transverse-snap	Weld	Lifting-bark	Desiccation-fissure	Frost-crack	Fluting	Ivy
Barbastelle																				
Bechstein's bat																				
Alcathoe's bat																				
Brandt's bat																				
Daubenton's bat																				
Whiskered bat																				
Natterer's bat																				
Leisler's bat																				
Noctule																				
Nathusius' pipistrelle																				
Common pipistrelle																				
Soprano pipistrelle																				
Brown long-eared bat																				
Lesser horseshoe-bat																				

Key: RL = *"reasonable likelihood"* (i.e. 'more-likely-than-not'); P = 'potential'.

In practice, atypical PRF situations are in any case rare, and programming the episodic memory will discipline the mind to shut out conjecture and focus solely on those PRFs for which there is tangible evidence of roosting occupation.

5.4 To survey or not to survey?

The outcome in the case of *Cheshire East Council v Rowland Homes* [2014] EWHC 3536 (Admin) and the Natural England EPSL Policy 4 criteria (Natural England 2016) suggest that even where the conclusion of the desk-study is that there are grounds to support a hypothesis that it is 'more-likely-than-not' that a tree-roost might be present, a survey may not be necessary.

The Policy 4 criteria suggest that the decision as to whether or not to survey should take into account two separate items, comprising:

1. Whether the ecological impacts of the development can be predicted with sufficient certainty;

   and

2. Whether mitigation or compensation will ensure that the licensed activity does not detrimentally affect the favourable conservation status of the local population[3] of any European Protected Species (Natural England 2016).

The spirit appears to be that to support the assertion that no survey is warranted, there should be sufficient evidence to demonstrate beyond reasonable doubt, that there will be no negative effect, regardless of the scenario.

The appraisal of the situation will therefore need to consider and balance the following three aspects:

» **Aspect 1** – the operation proposed.
» **Aspect 2** – the predicted ecological baseline.
» **Aspect 3** – the proven efficacy of the ameliorating options available (i.e. avoidance and mitigation).

Aspect 1

The details of the operation proposed will be specific to the circumstance, which are beyond the confines of this book to anticipate. However, considering guidance for EcIA published by the Chartered Institute of Ecology and Environmental Management (CIEEM 2016), the factors that would need to be unequivocally demonstrated would reasonably comprise:

» The extent, timing and duration of the operation.
» The frequency of the impact (i.e. whether the operation will encompass a series of the same impact).
» The magnitude of the impact.
» The degree of certainty that there will be an effect.
» The direction of the effect – positive or negative.

---

3   Defined as 'a group of individuals of the same species that live in a geographic area at the same time and are (potentially) interbreeding (i.e. sharing a common gene pool). This is an ecological definition, which includes the potential to interbreed, rather than simply meaning the collection of individuals of a species in a given area' (Natural England 2016). Note: Natural England have not yet defined local definitions of 'favourable conservation status'.

» If a negative effect is anticipated, its reversibility (i.e. whether any damage can be put right, and in what time-frame).

Aspect 2

Bearing in mind that at this stage the focus is upon existing information, the predicted ecological baseline will typically comprise a combination of either or both of the following:

» Positive evidence that tree-roosts have historically been present within the site, or outside the site but within the Zone of Influence[4];

and/or

» A 'structure-based'[5] assessment using habitat characteristics as a predictive indicator of the likelihood tree-roosts might be present. In the context of this book, that would be achieved by the performance of the intelligence-gathering as set out in Chapter 4 and the desk-study set out above.

Aspect 3

Finally, Aspect 3 will assess:

» Whether the species that might be present can be accommodated within the confines of the operation proposed.

» That derogation would not be detrimental to the maintenance of the populations of the species concerned at favourable conservation status within their natural range.

This will require the balance of the predicted impacts upon the predicted ecological baseline with the evidence that the impacts can be ameliorated to an acceptable level within a reasonable length of time. This balance should consider each of the impact assessment factors and seek ways to minimise their effects.

As the evidence of the efficacy of ameliorating action is changing all the time, the collation of this evidence will require a review of the evidence each time the appraisal is performed. However, in the context of compensation for the loss of individual tree-roost features, two publications are of particular value at this time:

» Poulton S 2006. *An Analysis of the Use of Bat Boxes in England, Wales and Ireland for The Vincent Wildlife Trust*. Vincent Wildlife Trust, County Galway, Ireland.

» Korsten E 2012. *Vleermuiskasten: Overzicht van toepassing, gebruik en succesfactoren*. Bureau Waardenburg BV, Culemborg.

If it is concluded that further data-collection would not be proportionate, it is recommended that the evidence that has led to this conclusion is reviewed by the Natural England Discretionary Advice Service (DAS). Experience has shown that the DAS offers a balanced assessment that is productive, even where the outcome is at odds with that hoped for.

If it is concluded that further data-collection is necessary, the desk-study will have identified where the knowledge-gap lies, and will therefore have identified the specific aim for the surveillance. However, the knowledge-gap may not necessarily be in respect

---

4   How old that evidence might be will depend upon the circumstances. For example, a roost within a lightning-strike on an oak that was last recorded ten years ago would be more likely to exist today than a roost in a woodpecker-hole on a crack-willow.

5   Structure-based indicators are stand-level and landscape-level (spatial) features that can be predicted to be of value to the target species. These include individual habitats and their structural complexity, plant species composition, connectivity and heterogeneity, etc. (Lindenmayer *et al.* 2000).

of the predicted ecological baseline, or the proven efficacy of the ameliorating options available.

If the details of the development proposed (i.e. the extent or layout) are uncertain, the only option available may be to anticipate the worst-case outcome in respect of the conclusions of the structure-based assessment (i.e. the prediction of the ecological baseline established by the desk-study analysis). Unless it can still be concluded that there are insufficient grounds to suggest a *"reasonable likelihood"* that the development would detrimentally affect the bat roosts potentially present, further action will be required in order to explore exactly what that effect might be.

5.4.1 Reality checkpoint

The variables that might impact upon tree-roosts are multitudinous, as are the accounts of the effective mitigation of impacts and compensation for residual effects. Although it would be great fun to attempt to anticipate the impacts of every variable that might be encountered, and pair them with effective strategies for the amelioration of their effects, this would detract from the message of this book, which is simply how to find tree-roosts.

Continuing with the threshold of what is 'more-likely-than-not', advice as to how to find tree-roosts will be supported by an evidence base that robustly demonstrates that:

» It is 'more-likely-than-not' the individual species of bat or evidence of their presence will be encountered if they are present.

» It is 'more-likely-than-not' that if bats are encountered, conclusive identification will be possible.

» It is 'more-likely-than-not' that a reliable count will be achieved.

» It is 'more-likely-than-not' that, where an aggregation is encountered, it will be possible to record their sex.

The remaining chapters of this book will therefore aim to provide sufficient information to inform a surveillance design that will be 'more-likely-than-not' to provide any missing or inadequate information and result in a data-set that will be sufficient to inform an impact assessment.

CHAPTER 6

Ground-Truthing

In this chapter	
Introduction	The objective of ground-truthing
Ground-truthing method	Timing, operation and equipment
Ground-truthing interpretation	The interpretation of the ground-truthing results and their collation into a format that has a practical application

6.1 Introduction

Ground-truthing is the first test of the hypothesis defined by the results of the desk-study. The objective of the ground-truthing is fourfold:

» Are the PRF types identified as 'more-likely-than-not' to be present, actually present?
» Where are they?
» At what height are they?
» Can they be accessed for close-inspection?

This is achieved by a structured search of the habitat for the PRFs that are 'more-likely-than-not' to be exploited, and also those that have the 'potential' to be exploited.

If such PRFs are encountered, all the potential roost trees present should be identified, marked, mapped, and the heights of their PRF recorded in order to inform the second test of *"reasonable likelihood"*.

The data gathered at this stage are then analysed to assess:

» Whether PRF present are in the height range exploited by the bat species in whose territorial range the site occurs.
» When each bat species is 'more-likely-than-not' to be present.
» In what numbers they are 'more-likely-than-not' to be encountered.

6.2 Ground-truthing method

6.2.1 Preparation

The ground-truthing is the fulcrum of tree-roost surveillance. It is what separates the field into those people who find tree-roosts and those people who do not, and do not learn. It is vital to the success of the operation that the ground-truthing is a targeted search.

This is not zone reconnaissance; that was performed at the intelligence-gathering stage. This is a targeted search focusing on specific features to pinpoint them, and begin to understand their potential value.

To do this, the mind must be programmed by scrutinising the Desk-study field summary table (Table 5.2) and mentally visualising the PRF types in the context of the specific tree species.

It is helpful to build-up a photographic reference library that can be used to create crib-sheets for each tree species, which may be kept in a vehicle for reference upon arrival at the site. As these crib-sheets represent tangible evidence that the surveyor is both disciplined and conscientious, they may also be produced if the survey result is challenged.

Time spent on preparation will result in effective programming that will reward disproportionately in terms of confidence – both the surveyor's self-confidence, and the confidence the result is accurate. Surveyors who are disciplined in their preparation rapidly advance to a level of skill that is such that they can pick out PRFs from a significant distance because they soon learn to recognise the environmental situations that are associated with PRF formation, and the structural character of the individual tree species that is associated with each PRF type.

6.2.2 Timing
There is one fundamental difference between good and bad ground-truthing: *timing*. In the vast proportion of circumstances, the window is limited to November–April[1] when: (i) the foliage is down; and (ii) light penetrates to the woodland floor.

The first aspect is obvious; leaves obstruct vision. In semi-natural woodland, even in winter, the trees themselves obstruct your view, but this can be overcome by retreating to get a better angle. When the trees are in leaf, retreating is not an option because the foliage on neighbouring trees in the canopy and shrub-layer obstruct the field of vision. In mid-summer, a multitude of other obstructions may confound the truthing, such as bracken *Pteridium aquilinum* and Himalayan balsam *Impatiens glandulifera*.

Even in softwood plantations and former hardwood timber crops (i.e. coppice with standards) the deformities in secondary 'weed' trees may hide PRFs from a distance. It is therefore vital that the field of vision is sufficient to pick out the shape of these trees so they are not overlooked.

In open habitat, the situation is not so very different as might be hoped. Typically, the trees are more spreading and many species have complicated growth forms and large leaves, such as horse chestnut *Aesculus hippocastanum*, sycamore *Acer pseudoplatanus* and common lime *Tilia × europaea*. It is possible in some situations to comprehensively inspect individual butchered trees in mid-summer on a hedgerow or in a park, but this is sufficiently uncommon to be considered atypical.

On a river bank, you need to walk both banks to see the PRFs over the water, and you simply cannot see them when the trees are in leaf.

Referring to Figure 6.1, the images speak for themselves.

But the foliage not only obstructs a clear field of view. Studies have shown that foliage may decrease light penetration 100-fold beneath the canopy (see Martin (1990) for a detailed review and discussion). The human eye takes approximately 45 minutes to adjust

1 Obviously, there are early and late winters and early and late springs (everything in nature is on a sliding scale), but a really good indicator that the mapping season is about to begin and is coming to an end is the horse chestnut: it drops its leaves early and it comes back into leaf on average three weeks before oak and beech, so when you see the horse chestnut in leaf, you know you need to finish any outstanding ground-truthing as soon as possible.

Figure 6.1 An example of the obstructive effect of foliage. The two images were taken from the same position – the image on the left was taken in July, that on the right in January.

to darkened conditions, but many surveyors fail to take into account that this period is extended in proportion to the amount of time spent in direct sunlight. As a result, if you drive to a site on a beautiful summer's day, and walk straight into a wood, you are not only constrained by the foliage, but you are also constrained by your failure to allow your eyes to adjust to the lower light conditions, and if you later walk into a glade, you will be back to square one and have to again wait under the canopy for your eyes to adjust.[2]

There is, however, one final aspect: the angle of the sun.

In the summer the sun has risen before most of us have started our working day, and is high above our heads for all our effective field-time. As we are attempting to search for PRFs from the ground to the canopy, the position of the sun is in our eyes and is therefore a significant constraint. In the winter, the sun rises much later, and remains offset to our line of sight. Therefore, providing we are not staring into it, the sun is a significant benefit because it illuminates the tree(s) from one side, and lights up the whole of the structure equally. It also throws any darkened cavity or fissure into sharp relief.

Whilst in some harsh winters November and April are acceptable, if you map in the period December–March, the constraints in respect of foliage, ocular adjustment, light penetration and angle are all removed and significantly reversed, to a positive situation where mapping is at its most effective, quickest, and therefore most cost efficient.[3]

An alternative that might be considered when there are timing constraints might be to climb every tree in the site, and an argument that is sometimes extended in support of this course of action is that some PRFs cannot be seen from the ground.

There are situations where climbing might be considered in tandem with ground-truthing, and one such situation is set out in the following subsection. However, in any situation where climbing is considered in addition to, or instead of, ground-truthing, it should be supported by a proportionality test (such as that set out at the close of Chapter 1).

---

2   Anyone who has ever watched a war film set in a jungle situation will have noticed that ambushes always take place when one side walks through open habitat and then enters under the canopy without stopping. The ambush is successful because the eyes of the force lying in wait have fully adjusted to the conditions, and the eyes of the moving force have not, and are squinting into effective darkness. PRFs are every bit as well camouflaged as an ambush; think about it …

3   Another significant improvement is that the herb-layer is at its least irritating (i.e. the stinging nettles *Urtica dioica* and bramble *Rubus fruticosus* agg. are down).

6.2.3 The operation

A map of the habitat will have been produced at the intelligence-gathering stage. The data now needed will be gained by performing a 360° inspection of each tree.

As PRFs do not typically occur as isolated individuals but in pockets and lines, it is therefore sensible to start by performing a circuit of a wood to look for pockets of damage and moribund and dead standing timber.[4]

Experience has shown that while most sites can be physically searched by an individual surveyor, in most situations, two people working together – one acting as scribe and the other taking all the measurements and photographs – is cost effective due to the increase in the speed of recording. This is particularly pertinent to hedgerow and riparian fringe as both sides can be searched in one movement without the need for continuously back-tracking.

In sites with year-round obstructions, such as cherry laurel *Prunus laurocerasus* or rhododendron *Rhododendron ponticum*, it is sometimes least infuriating for cluttered areas to be searched by three people, with one remaining outside the clutter and giving directions and taking notes while the other two push through from opposite sides.

In cluttered habitat (such as carr woodland) and any situation that encompasses steep slopes, there are two transects that have been found effective:

» The fan (see Figure 6.2 – the fan is useful in small areas of cluttered habitat).

» The box (see Figure 6.3 – the box is particularly useful on slopes).

Figure 6.2 The fan transect, used to ground-truth two areas of cluttered woodland.

---

4 Satellite imagery may also be used for identifying dead trees.

Figure 6.3 The box transect, used to ground-truth an east-facing ravine-side.

Finally, it should be noted that ground-truthing is not equally effective in all habitats or between tree species. In a mature beech *Fagus sylvatica* woodland, ground-truthing is unlikely to reflect the true situation due to the size and structure of the trees and the frequency of bark-stripping by squirrels that may result in a relatively high number of wounded limbs hollowing like dug-out canoes, with a narrow entrance that is only visible from above. However, even in relatively open habitat the ground-truthing will not identify every single one of the PRFs present. Radiotracking studies have demonstrated that even where a roost-tree has been identified, the actual roost feature may not be visible from the ground (e.g. Murphy *et al.* 2012).

Ultimately, ground-truthing (as the name suggests) comes down to honesty: if you are confident that you have the measure of the habitat and that it is 'more-likely-than-not' that you have a robust map, then you will be able to achieve the objective. On the other hand, if you have reservations about whether the map is a robust reflection of the PRFs present, then the question has been answered in the negative.

In such a situation, ways confidence might be improved should be readily apparent. In the case of the beech example above, it might be possible to gain sufficient confidence by performing some targeted climbing of trees at equidistant spacing (say every 100 m) to a height from which the climber can scan the surrounding trees over a wide radius, in order to direct a recording-team on the ground to trees that hold PRFs that would otherwise be invisible to them. It should be noted that this operation would only be likely to be sufficiently effective if performed when the foliage was down.

6.2.4 Equipment
To perform ground-truthing you will need:

» A recording form and a pen, or an electronic recording medium such as a tablet.

» Binoculars.

» Hazard-tape/forestry marking paint/tree-tags.

Table 6.1 Ground-truthing recording form

1st Recorder:		2nd Recorder:
GROUNDSMAN	**SITE NAME**	
	DATE	
	HABITAT (Phase 1)	
	GRID REFERENCE (OS not lat/lon)	
	Tag No. / Tree Ref.	
	TREE SPECIES	
	TREE ALIVE/DEAD	
	DBH (Diameter at Breast Height)	
	TREE HEIGHT	
	PRF on STEM/LIMB	
	PRF TYPE	
	DIRECTION PRF FACES	
	PRF HEIGHT (approximate)	
	COMMENTS	**Do not forget the photographs!**
CONTEXTUAL PHOTO		**PRF CLOSE-UP PHOTO**
insert		*insert*

» A hand-held GPS.

» A DBH tape/tree calipers.

» A clinometer.

» A compass.

» A camera.

This list is non-negotiable: whether you are assessing one tree or a hundred, to perform accurate ground-truthing sufficient to charge a fee, this is the equipment you will need, and you must be confident in its use.

Table 6.2 Notes on certain tree species which have structural idiosyncrasies that make the PRFs they hold difficult to detect if the search is not targeted at the right part of the tree, and at the right time of year

Tree species	Structural idiosyncrasies
Field maple	Hides wound PRFs in forks and on the upper side of limbs
Sycamore	Attacked by squirrels bark-stripping and may form a series of wound PRFs from the butt into the canopy, many of which are hidden within the deformed structure of the tree, and when foliage is out, by twig growth and large leaves
Hazel	Wound PRFs frequently occur between two coppice stems that cross each-other but have not 'welded'
Beech	Attacked by squirrels bark-stripping whilst sitting on the upper side of branches. As a result, beech forms the greater proportion of wound PRFs in positions that are invisible from the ground
Ash	Entrances into scab-type canker PRFs can be particularly hard to find, and tricky to get an endoscope into. A low-diameter endoscope snake will be required for most ash cankers
Domestic apple	Hides wound PRFs in forks
Pedunculate oak	Puts on small burr-growths of twigs on the woundwood seams of hazard-beams. When the foliage is out, these growths mean that the PRF may be missed, even when its location is known
Lime	Puts on large burr-growths around wound PRF. When the foliage is out, these growths mean that the PRF may be functionally invisible, even when its location is known

Recording form

A template recording form is provided at Table 6.1. This form is the 'groundsman' section of the overall BTHK PRF recording form[5] (a Word copy of this table can be downloaded from www.battreehabitatkey.com).

Regardless of the recording medium, a record should be made of all the PRFs on the tree from top down. The BTHK Database assigns an alphanumeric code to PRFs on the same tree working from top down: 1a, 1b and 1c, with these representing the highest, middle and lowest PRFs. If the PRF has more than one entrance, both entrances will require recording, but this will be dealt with in the following chapter.

Binoculars

Do not spend too much time with the glasses clamped to your eyes. You will pick up more by scanning with the naked eye. You are looking for structural clues in the trees' shape, not searching every inch for holes. Just use the binoculars to assess whether the PRF is open or discontinuous, and to identify its form.

In fact, each tree species has subtle idiosyncrasies in the position of the PRFs they form, and having a broad idea of these will inform more effective and accurate mapping. Table 6.2 sets out the situation in respect of certain 'tricky' tree species.

Hazard-tape/forestry marking paint/tree-tags

These are used to mark the tree so it can be refound if repeat surveillance is required.

Hazard-tape has two advantages: it is temporary; and you can write on the tape. However, it may slip down the tree or get torn off by members of the public.

Forestry marking paint is permanent and can be seen from a great distance, but only use pink as it denotes a survey mark. Yellow means cut down the tree, and orange means the tree is diseased, which it may well be, but the next step from orange is yellow on a

---

5   The BTHK PRF recording form is a variation of an original data-collection format defined by Sedgeley and O'Donnell (1999) and used very effectively by Ruczyński and Bogdanowicz (2005).

large scale! If you draw attention to disease, you run the risk of losing all your PRF trees in one sweep. Spray has the added advantage that it can be used as a dot and by marking all the stems on the same side aggregations can be seen from a distance when approached at the right angle, or the stem can be ringed with spray, thereby meaning it will be easily found when approached from any angle.

Tree-tags are unobtrusive and provide a long-term marker. They are, however, relatively small and may be obscured by vegetation in the summer months.

GPS
A good-quality GPS is vital if an accurate map is to be achieved, and if the trees are to be refound, but be aware that these devices take a good 10–15 minutes to settle and should be kept absolutely still while they are finding satellites. It is therefore sensible to switch on the unit upon arrival at the site. Always take a full ten-digit grid reference. This is another reason for mapping in winter when the foliage is down – a more accurate location is plotted and therefore a more accurate map obtained.[6]

DBH tape or tree calipers
The only way to get an accurate diameter is a DBH tape or calipers. A DBH tape can be carried in a pocket but Haglöf tree-calipers (available from nhbs.com) will save a great deal of swearing when recording trees on river banks and in thorn hedges, or where there is barbed-wire.

Clinometer
The only way to get an accurate tree height is with a clinometer and *this datum is vital for the ground-truthing analysis*. In addition, an accurate height will be required when deciding how the PRF will be accessed for inspection.

Compass
The aspect of the PRF is required in order that the PRF may be found again when the tree is in leaf. This is particularly important if the climbing will not be performed by the surveyor who performed the ground-truthing.

Camera
At least two photographs are required: one of the entire tree and the PRF in one frame; and a close-up of the PRF so there is some chance of finding it when the tree is in leaf. These two photographs are also the minimum requirement required if the PRF record is to be accepted for inclusion on the BTHK Database.

6.3 Ground-truthing interpretation
Assuming PRFs are present, the second test of *"reasonable likelihood"* is performed by comparing the height of each PRF to see if it is within the range of roost records on the BTHK Database, and also establishes the vitality of the tree. This serves three functions:

1. It demonstrates whether the PRF is within *all* the required parameters.
2. It informs when the PRF should be surveyed and directs action to the season it is 'more-likely-than-not' that bats will be present.
3. It identifies whether the PRF is accessible by virtue of its height, and the tree's vitality, and therefore which surveillance methods might be applied.

---

6   Always interrogate any third-party GIS map as these typically do not have gradient lines, and what looks like a level walk in the park can in reality be a Herculean yomp over difficult terrain. In particular, tree canopies have a levelling quality in satellite images, and can disguise deep ravines and steep mounds.

Table 6.3 Ground-truthing account template

SITE NAME		
TREE REFERENCE/TAG No.		
OS GRID REFERENCE		
TREE SPECIES		
PRF TYPE		SATISFIES STAGE 1 CRITERIA? Y/N
PRF HEIGHT		SATISFIES STAGE 2 CRITERIA? Y/N

BAT SPECIES	Predicted presence (RL/P*)	Winter		Transitory (spring-flux)		Maternity — Pregnancy		Maternity — Nursery		Mating		Transitory (autumn-flux)	
		Jan	Feb	Mar	Apr	May	Jun	Jul	Aug	Sep	Oct	Nov	Dec
Barbastelle													
Bechstein's bat		▓	▓	▓									▓
Alcathoe's bat		▓	▓	▓									▓
Brandt's bat		▓	▓	▓				▓	▓	▓	▓	▓	▓
Daubenton's bat		▓	▓										▓
Natterer's bat													
Whiskered bat		▓	▓	▓				▓	▓	▓	▓	▓	▓
Leisler's bat													
Noctule													
Nathusius pipistrelle						▓	▓	▓					
Common pipistrelle													
Soprano pipistrelle													
Brown long-eared bat													
Lesser horseshoe-bat		▓	▓	▓	▓	▓							

VITALITY (alive/moribund/dead)	
ACCESS POSSIBLE (Y/N)?	
ROOST RECORDS HELD ON BTHK DATABASE COMPARABLE TO THIS CONTEXT (Y/N)? Identify the roost position. List any field-signs associated with bat presence and any competitors	

*RL = *"reasonable likelihood"*; P = potential.

The interpretation comprises five stages and may be performed using the template provided in Table 6.3, proceeding as follows:

Stage 1

Are the PRF types identified as 'more-likely-than-not' to be present, actually present?

» Yes: Proceed to Stage 2.

» No: No further action required.

Stage 2

Using the PRF accounts provided in Chapter 3, do the PRFs present fall within the height ranges known to be exploited by the bat species that are 'more-likely-than-not' to occur within the site?

» Yes: The second test of *"reasonable likelihood"* is satisfied; highlight the relevant species in the left 'Bat Species' column on the ground-truthing recording form (Table 6.3; a Word copy of this table can be downloaded from www.battreehabitatkey.com) and proceed to Stage 3.

» No: The justification for proceeding should be questioned.

Stage 3

Identify the periods in which the individual bat species might be present using the summary provided in Table 2.3 in Chapter 2 and the individual PRF accounts in Chapter 3. This information may be used to determine the surveillance intensity applicable to each of the individual PRFs present. Surveillance intensity is rarely the same for all the PRFs present: due to the different seasonal exploitation of different PRFs by different bat species, there is no 'one-size-fits-all' prescription.

Stage 4

Consider the height of the PRF and whether it can be accessed for close-inspection. As a broad guide:

1. PRFs up to 2 m will be accessible for inspection from the ground.
2. PRFs up to 6 m will be accessible from a ladder.
3. PRFs at heights above 6 m will require specialist access methods, such as the use of a mechanical elevating work platform (MEWP) or arboreal climbing.

In most cases PRFs on living trees will be accessible by arboreal climbing methods. There are, however, situations where attempting access would be reckless due to the weakness inherent in the structure. Remember, a survey licence is not a licence to take risks: if the survey team accidentally damage or destroy a roost they are still guilty of an offence. Furthermore, careful consideration of access feasibility will be required if the tree is dead, or the PRF is in a situation where a MEWP would be required to safely access it, but the habitat and/or terrain preclude its use. In the latter situation, access may be impossible and it may therefore be necessary to resort to remote-observation or static-netting.

Stage 5

Records held on the Bat Tree Habitat Key Database typically have a ten-digit OS grid reference. This means that the records can be viewed in landscape context on websites such as www.gridreferencefinder.com.

Stage 5 of the analysis is therefore to see whether the tree species, habitat and PRFs present on the site accord with the context of records held on the Database. This is simply done by filtering the Database using the tree species, habitat and PRFs, and comparing the truthing result with the Database. If matching records are found, the grid reference and contextual photographs are requested via www.battreehabitatkey.com and the environments can be compared to see how similar the context may be. If the context matches, then the Database can be used to identify the bat species present and numbers, the typical roost position (i.e. above, below or to the side of the entrance), when bats have been recorded in that context, and what field signs and competitors were associated.

By being disciplined and completing a Ground-truthing recording form for each PRF, the episodic memory is programmed and a thorough record is compiled of the PRFs within a given area of wooded habitat. The surveyor very soon notices the pattern of PRF types and their individual forms within different variations of the same vegetation community, and discerns distributional bias. In addition, a 'toolbox' account of the situation is compiled that can be provided to any third party that may be involved with subsequent surveillance; for example, a climbing team may use the accounts to provide a fixed quote.

Regardless, a map of the PRFs has now been achieved and there is a robust account of the potential bat-roost interest within that particular site that can be used to inform further surveillance action and choose an appropriate surveillance method.

CHAPTER 7

Choosing the Surveillance Methods

7.1 Introduction

In broad terms, surveillance can be divided into four operations:

» Close-inspection.

» Remote-observation.

» Static-netting.

» Radiotracking.

As with everything in nature; there is no 'one-size-fits-all' surveillance method; all have strengths and weaknesses. However, as this chapter will demonstrate, the weakness of one can be overcome by combining it with another.

This chapter provides a summary review of the four surveillance methods that will lead into more detailed descriptions of the methods in the following chapters.

7.2 Close-inspection

7.2.1 Rationale

Close-inspection comprises a visual inspection of the ground, tree and both exterior and interior of the PRF using a combination of cree-torch and endoscope.

7.2.2 Strengths and weaknesses

Where close-inspection is reasonably practicable it should be the foundation of any surveillance campaign because:

» It is the only method that can inform an appraisal of suitability.

» It is the only method that can be repeated in all seasons.

» It is the only method that can detect field-signs.

» It is the only method that will reliably detect the presence and absence of competitors.

Ultimately, close-inspection is the only method that can establish the suitability of an unoccupied PRF and detect 'change', and therefore the only method that builds data in the absence of bats, and builds confidence in a negative the more visits are made. Close-inspection is also the only method that is entirely reliable in winter and both flux periods.

However, close-inspection is the most invasive of the four methods available; it will certainly disturb any bats present and the method has the potential both to injure bats and damage or destroy the roost feature.

Even where the method is proficiently applied, it is far from perfect and has many constraints: comprehensive inspection may not be possible; the field-signs have associated false-positives as well as false-negatives; and where aggregations of bats are encountered it may not be possible to confidently identify them, infer the sex or even count them. Finally, due to roost-switching and fission/fusion behaviour, the probability of encountering colonies of several species builds at a relatively low rate during the pregnancy and nursery periods (c. 5–7.5%), and only builds at all where visits are made sequentially.[1]

Table 7.1 gives a summary of the strengths and weaknesses of close-inspection.

Chapter 8 sets out the close-inspection method.

7.3 Remote-observation

7.3.1 Rationale
Remote-observation comprises fixed observation of the PRF from a distance using a zero-lux video-camera or an equivalent night-vision device to see if any bats emerge.

7.3.2 Strengths and weaknesses
Remote-observation may be appropriate where close-inspection is impossible or not reasonably practicable, or the presence of bats has been established by close-inspection but count data is still required.

The method's greatest strength is in combination with close-inspection, static-netting and/or radiotracking, particularly where count data is required. Although it may detect competitors that may suggest the PRF is suitable for exploitation by roosting bats, its weaknesses in isolation are significant in that it cannot detect change in the internal environment, can only be conclusive in the positive, and may not reward with conclusive identification of the species or inference of the sex.

Not to be overestimated: the method is mind-numbingly boring and requires a particular kind of discipline if attention is not to wander. In addition, the lux levels many species emerge in are below the levels at which the human eye can see. As a result, the combination of attention span and darkness mean the method is only reliable where it is performed using a zero-lux camera or an equivalent night-vision device that can record footage allowing later review, and the method is labour intensive in the time it takes both to perform the survey and review footage.

---

1 In practice, this is often less of a constraint for close-inspection than for remote-observation and netting, because field-signs such as droppings will often be sufficient to get the identification through DNA analysis, and count data can be inferred from scientific accounts (basically, the typical size of colonies and sub-groups of most species has already been demonstrated, and the combination of the PRF type, season and number of droppings can be combined to inform a test of *"reasonable likelihood"*).

Table 7.1 Close-inspection: strengths and weaknesses

Seasonal efficacy	Winter (Jan/Feb)	Spring-flux (Mar/Apr)	Pregnancy (May/Jun)	Nursery (Jul/Aug)	Mating (Sep/Oct)	Autumn-flux (Nov/Dec)
	Yes	Yes	Yes	Yes	Yes	Yes
Suitability and environmental cues	Yes: both					
Field-signs	Yes: all					
STRENGTHS	• Will inform an assessment of suitability at the outset of a surveillance campaign, allowing effort to be targeted to the bat species proven to exploit the feature, and the season for which a *"reasonable likelihood"* of presence has been demonstrated. • The only method that will establish winter usage. • Gives confidence in a negative result for that particular visit, and a series of visits, thereby demonstrating where the colony 'is not' as well as where it is. • May yield sufficient field-signs to demonstrate, or at least infer, roost status even in the absence of the bats themselves. • May identify mixed-species roosting (i.e. two species occupying the same roost on the same day). • May identify the presence of a competitor that may preclude roost presence. • Several trees may be surveyed by a single climbing team of two or using a mechanical elevating work platform (MEWP) in a single day. Where it is successfully used in isolation, close-inspection is therefore typically the most cost effective of the four methods.					
WEAKNESSES	• Where there is evidence that bats are present, close-inspections should only be performed by a surveyor holding a Level 2 licence or above. • It is the most invasive method of the four available and has the potential to injure bats and damage the PRF. • The PRF must be accessible. • It must be possible to comprehensively search the internal areas. • Environmental cues and field-signs should be carefully interpreted, acknowledging the potential for false-positives as well as false-negatives. • Where bats are present, in the absence of droppings, identification is not always possible. • An exact count of the bats or even a rough estimate may not be possible. • The sex of the bats cannot be inferred from group size in many situations (i.e. Daubenton's bat, Natterer's bat, Leisler's bat, noctule and lesser horseshoe-bat). • Due to roost-switching and fission/fusion behaviour, the probability of encountering colonies of several species builds at a relatively low rate during the pregnancy and nursery periods (c. 5–7.5%), and only builds at all where visits are made sequentially.					

Note: Close-inspection is not appropriate where maternity colonies are already known to be present and would in any case be unlikely to reward with any information that could not be inferred by an inspection soon after the colony had moved on.

As with close-inspection, due to roost-switching and fission/fusion behaviour, the probability of encountering colonies of several species builds at a relatively low rate during the pregnancy and nursery periods (*c.* 5–7.5%), and only builds at all where visits are made sequentially.

Table 7.2 gives a summary of the strengths and weaknesses of remote-observation. Chapter 9 sets out the remote-observation method.

Table 7.2 Remote-observation: strengths and weaknesses

Seasonal efficacy	Winter (Jan/Feb)	Spring-flux (Mar/Apr)	Pregnancy (May/Jun)	Nursery (Jul/Aug)	Mating (Sep/Oct)	Autumn-flux (Nov/Dec)
	No	Sometimes	Yes	Yes	Yes	Sometimes
Suitability and environmental cues	No: neither					
Field-signs	Yes: competitors only					
STRENGTHS	• It does not require a survey licence. • The least invasive method of the four available, and has no potential to injure bats or damage the PRF. Furthermore, when performed sensitively remote-observation results in the least disruption to any bats present. • Can be effectively performed as an all-night observation prior to felling. • The PRF does not have to be accessible. • Will give a reliable count. • May identify mixed-species roosting (i.e. two species occupying the same roost on the same day). • May identify occupation by a competitor that may suggest suitability or preclude roost presence.					
WEAKNESSES	• Will not inform an assessment of suitability. • Will not establish winter usage. • Can only record a positive; cannot give confidence on a negative result as bats may be present but may not emerge. • Will not detect field-signs such as odour, substrate cues or droppings. • The PRF must be clearly visible. • Requires the use of electrical equipment that must very carefully deployed if it is to yield useful data and a reliable result. • May not be able to conclusively identify the bats to species (in particular, *Myotis* sp.). • May not be able to infer the sex of the bat(s). • May not be able to infer breeding condition. • It is the single most boring and soul-destroying survey method. • Is labour intensive in terms of survey and analysis and is equal to static-netting in terms of cost. • Due to roost-switching and fission/fusion behaviour, the probability of encountering colonies of several species builds at a relatively low rate during the pregnancy and nursery periods (*c.* 5–7.5%), and only builds at all where visits are made sequentially.					

7.4 Static-netting

7.4.1 Rationale
Static-netting comprises the physical deployment of an appropriate net by hand over the entrance of the PRF to catch any bats that might emerge.

7.4.2 Strengths and weaknesses
As with remote-observation, static-netting may be appropriate where close-inspection is impossible or not reasonably practicable. The immediate advantage of the method over remote-observation is that it provides not only count data, but allows conclusive identification and sexing, and even an assessment of breeding condition.

The method's greatest strength is in combination with close-inspection, remote-observation or radiotracking, in particular where species identification, sex and breeding condition is required. However, its weaknesses in isolation are significant in that it cannot detect change and is only ever conclusive in the positive. Furthermore, it is invasive and requires a high level of stamina and also skill if physical harm to any bats captured and potential damage to the PRF are to be avoided. As captures need to be retrieved, it is most effectively performed by a pair of surveyors equipped with a night-vision camera in order that the count is reliable.

It cannot be used at dawn to detect returns.

Finally, as with close-inspection and remote-observation, due to roost-switching and fission/fusion behaviour, the probability of encountering colonies of several species builds at a relatively low rate during the pregnancy and nursery periods (c. 5–7.5%), and only builds at all where visits are made sequentially.

Table 7.3 gives a summary of the strengths and weaknesses of static-netting

Chapter 10 sets out the static-netting method.

7.5 Radiotracking

7.5.1 Rationale
Radiotracking comprises the capture of free-flying bats with the subsequent attachment of a radiotransmitter that enables the bat to be located by pursuit by an individual surveyor, or triangulation by two or more surveyors using radioreceivers.

7.5.2 Strengths and weaknesses
The immediate advantage of the method over the other three is that it is guaranteed to find roosts, but is non-invasive and only disturbs a single bat during capture and the subsequent attachment of the radio-tag. Furthermore, as the bat is chosen from the outset, the method is ideal for targeting the species and colonies that might be most at risk from development and thereby informing a robust impact assessment. Finally, trapping and radiotracking remove the bias of encounter probability entirely, and are therefore particularly useful for the Annex II species: barbastelle and Bechstein's bat, as well as Alcathoe's bat. Trapping in isolation would also yield useful presence/absence data in respect of all other woodland species, but in particular Brandt's bat and whiskered bat (conclusive identification) and brown long-eared bat and lesser horseshoe-bat (presence and relative abundance).

However, its weaknesses in isolation are significant in that it can only identify the roosts that are occupied during the tracking period, and potentially only by a sub-group of a larger colony. As a result, it is often wise to combine the method with either close-inspection or remote-observation methods, particularly where a colony is of a species known to divide into smaller subgroups (such as the barbastelle and Bechstein's bat (see Greenaway 2008; Dietz and Pir 2011)).

Table 7.3 Static-netting: strengths and weaknesses

Seasonal efficacy	Winter (Jan/Feb)	Spring-flux (Mar/Apr)	Pregnancy (May/Jun)	Nursery (Jul/Aug)	Mating (Sep/Oct)	Autumn-flux (Nov/Dec)
	No	Sometimes	Yes (but see text*)	Yes (but see text*)	Yes	No
Suitability and environmental cues	No: neither					
Field-signs	No: none					
STRENGTHS	• Will give a reliable count. • Enables a conclusive identification to be made in the hand or supported by DNA from fresh droppings. • Enables conclusive sexing. • Enables an assessment of breeding condition. • Reliably identifies mixed-species roosting (i.e. two species occupying the same roost on the same day).					
WEAKNESSES	• Can only be performed on a Level 2 licence or above and any sort of trap other than a static-net will require a higher-level licence. • It is the second most invasive method of the four available and has the potential to injure bats and, although only rarely, damage the PRF. • Cannot be used for a dawn-return observation prior to felling. • Will not inform an assessment of suitability. • Will not establish winter usage. • Can only record a positive; cannot give confidence on a negative result as bats may be present but may not emerge. • Will not detect field-signs such as odour, substrate cues or droppings. • The PRF must be sufficiently accessible for the net to be used with confidence that bats will not be harmed and there will be no damage to the PRF. • *It should not be used when there is any possibility heavily pregnant bats or bats with dependent young might be present; typically, June through mid-July (see Collins 2016). • Static-netting requires a *very* high level of skill and stamina. • Can only record a positive; cannot give confidence on a negative result as the bats may have emerged late or not at all. • It is labour intensive in terms of survey and analysis and is equal to remote-observation in terms of cost. • Due to roost-switching and fission/fusion behaviour, the probability of encountering colonies of several species builds at a relatively low rate during the pregnancy and nursery periods (c. 5–7.5%), and only builds at all where visits are made sequentially.					

Note: Common sense should be applied in the use of a static-net. Where there is any potential for injury to the bat or damage to the PRF, the method should not be applied. Therefore, in the greater percentage of situations it will only be effective where the PRF entrance is sufficiently open for the net to be deployed and retrieved in one smooth movement (bearing in mind that the latter will be done in the dark).

Table 7.4 gives a summary of the strengths and weaknesses of radiotracking.

Table 7.4 Radiotracking: strengths and weaknesses

Seasonal efficacy	Winter (Jan/Feb)	Spring-flux (Mar/Apr)	Pregnancy (May/Jun)	Nursery (Jul/Aug)	Mating (Sep/Oct)	Autumn-flux (Nov/Dec)
	No	Sometimes	Yes (but see text*)	Yes (but see text*)	Yes	Sometimes
Suitability and environmental cues	No: neither					
Field-signs	No: none					
STRENGTHS	• It is not invasive with respect to the roost, and only an individual bat of a colony is disturbed for a short period when it is trapped and the radio-tag is attached. • The PRF can be inaccessible; the method will still detect the presence of the radio-tag in the tree. • Guaranteed to find roosts. • Encounter probability is entirely removed as a constraint. • The identity, sex and breeding condition of the bat is specifically chosen and therefore known from the outset. • Can survey a large area in a relatively short period, and is particularly useful in respect of species that occupy dead trees or large features in dead limbs, or roost in cryptic features in the high canopy that would be impossible/hazardous to climb, or might be damaged by close-inspection, i.e. barbastelle, Alcathoe's bat and Leisler's bat. • It is exhilarating and rewards the practitioner with the deepest understanding of the species.					
WEAKNESSES	• Requires a higher-level licence and the number of practitioners qualified to offer the method professionally is limited. • Will not establish winter usage. • Will not inform whether a PRF is not a roost; it may be suitable but not occupied that year, or during the tracking period, or by another species that was not tracked. • Will not detect field-signs such as odour, substrate cues or droppings. • In isolation, will not identify mixed-species roosting. • In isolation, will not identify the presence of competitors. • It is significantly disturbing to the bat(s) trapped and tracked. • *It should not be used when there is any possibility heavily pregnant bats or bats with dependent young might be present; typically, June through mid-July (see Collins 2016). • Roost presence is limited to the species that is tracked. However, the initial trapping does typically identify the presence of tree-species generally, in addition to those that are to be the subject of the trapping. • Misleading data must be 'factored-out' of results; in particular, the so-called 'panic-roosting' behaviour, where there is the potential that the tagged bat is occupying a roost purely because it was the nearest shelter available and the bat is traumatised by its capture and subsequent tag fixing. • Will not give a count of bats in the roost, merely where the tagged bat is roosting, it will therefore be necessary to perform remote-observation or netting to get a count. • Will not tell you if the tagged bat is sharing the roost with another species. • Requires a truly massive amount of experience before sufficient skill is achieved to perform it effectively (radiotracking might reasonably thought of as a 'lifestyle' rather than a profession). In addition, it is terrifically labour intensive and requires expensive equipment that needs regular maintenance.					

This book will go no further in discussing trapping and radiotracking. Readers are directed to the text describing the method written by Ian Davidson-Watts and included in the Bat Conservation Trust's *Bat Survey Guidelines for Professional Ecologists – Good Practice Guidelines* (Collins 2016).

7.6 Putting it all together

Table 7.5 provides a summary of the data that each of the four surveillance methods will reliably collect.

Table 7.5 The data that close-inspection, remote-observation, static-netting and radiotracking will collect (Y = yes; S = sometimes; N = no)

Method	Data required							
	PRF suitability	Environment change	Field-signs	Species identification	Count	Sex	Colony location	Competitor presence
Close-inspection	Y	Y	Y	S	S	S (inf)	Y	Y
Remote-observation	N	N	N	S	Y	S (inf)	S	S
Static-netting	N	N	N	Y	Y	Y	S	S*
Radiotracking	N	N	N	Y	N	Y	Y	N

* With potentially shocking results: a grey squirrel in a net does wonders for firing adrenaline and focusing attention.

It can immediately be seen that to gain a comprehensive appraisal, a combination of two or more methods is sometimes required.

Regardless, the suite of methods chosen should, in combination, be effective for *all seasons* with an >50% confidence threshold.

When deciding on a suite of surveillance methods, it is therefore vital that the objective is defined. Some questions that might define the objective are:

» Do suitable PRFs exist within the site?

» Do PRFs of a type known to be exploited by a specific bat species or suite of bat species exist within the site?

» Does a maternity colony of a species roost within the site?

» Is the site exploited by mating bats?

» Is the site exploited by hibernating bats?[2]

This will decide which method you use.

---

2   This is an often-overlooked aspect of roost surveillance. The fact is that the greatest value of some woodlands is as a hibernaculum. The situation is no different from a large subterranean system with individual bats roosting in discrete crevices or in groups on the walls or ceiling. Woodland bats are exactly the same, and may occupy a single compartment with individual bats each in their own crevice-type PRF in individual trees and groups of some species in the domed apex of void PRF. There is the potential to do massive damage, resulting in a significant negative effect if insufficient winter work is performed.

7.7 The proportionality test

When choosing surveillance methods, honesty is paramount, and experience has shown that it is often helpful to apply the Proportionality Test criteria in reverse order as follows:

4. Is the outlay proposed reasonable, i.e. do the ends justify the means and suit the circumstances – number of bats, rarity of the species, conservation trend, etc.?
Is close-inspection really the most cost-effective option? Is it 'more-likely-than-not' that barbastelle, Bechstein's bat or Alcathoe's bat might be roosting in the site?

If a colony is encountered during a close-inspection, but the species cannot be conclusively identified and/or a count is impossible, what will the fallback position be if it is 'more-likely-than-not' that the colony is rare or uncommon? Has the potential for this situation been anticipated?

3. Is the method and level of survey effort necessary to achieve the aim and is there no cheaper way of doing it?
Will the presence of any single species be a 'show-stopper'? If so, is a general survey appropriate?

Is the habitat too complex? Is it 'more-likely-than-not' that PRFs might occur in situations that are invisible from the ground (such as wound PRFs on the upper surface of beech limbs or behind ivy), or is the habitat too dense (such as situations with rhododendron, cherry laurel, bramble, etc.) so confident truthing cannot be satisfactorily completed? Are there trees holding PRFs in inaccessible areas?

Is there a superabundance of PRFs at significant height that would make close-inspection, remote-observation and static-netting more expensive than radiotracking?

Is there a superabundance of PRFs on dead or dying trees?

The list goes on …

2. Can I *prove* that the method is suitable to achieve the aim?
Do you have evidence that the suite of methods proposed are demonstrably effective to achieve the objective? If there is any doubt, there is no doubt; there are no grey areas – either the method will produce data that can be interpreted, or there is an unacceptable element of gambling.

Ask yourself: *is this the most effective method there is, or just the most effective I personally can do?*

If there is any uncertainty, then it is wise to keep in mind Criterion iii of the Chartered Institute of Ecology and Environmental Management (CIEEM) Code of Professional Conduct (CoPC):

> *'As a member of CIEEM I shall: seek advice and assistance if I am involved in topics outwith my sphere of competence.'* (CIEEM 2013).

The sphere of competence is not, however, limited to competence in respect of a particular method, it may also be in respect of an individual species in a wider suite of species, or in the context of a particular habitat type or geographic location. Even where the team is confident that they have an effective method, it may be sensible to seek a second opinion where there is the potential for public scrutiny. This might take the form of commissioning a peer-review from another consultancy, or by using the Natural England Discretionary Advice Service.

Remembering that all development proposals involving woodland have the potential for public outcry, if the project has the potential to be scrutinised at a Public Inquiry, a second opinion will (if nothing else) demonstrate integrity in line with Criterion vi of the CIEEM CoPC:

'As a member of CIEEM I shall: uphold professional integrity whilst maintaining the highest standards of ethical conduct.' (CIEEM 2013).

1. What is the specific and identifiable aim for the survey proposed?

When the method has been defined, an interpretation framework will be required into which the data can be entered and this will require a specific and identifiable aim from the outset: *a threshold should be defined that will trigger action,* such as the conclusion of the surveillance, a change in the primary method, an increase/decrease in intensity, an extension of time.

All too often, surveillance programmes are run for a specific period, regardless of the data accrued. Again, it is sensible to ask whether the presence of any species will be a 'show-stopper' in order that resources are not unnecessarily expended if the viability of the project is already in doubt.

Having established the surveillance aim, it is vital that everyone involved with the surveillance is aware of the threshold in order that focus is maintained, the right data are recorded, and the data are collated and reviewed after each visit so as to inform any future visit. By so doing, the data will be correctly filed (rather than as a big muddle to be unravelled after a given number of visits), and the programme will be concluded the moment the threshold is reached.

CHAPTER 8

Close-Inspection

In this chapter	
Introduction	The data the method is to collect
Equipment	The equipment needed to collect the data
Health and Safety	Considerations in respect of biotic hazards
Close-inspection method	How to collect and record the data
Interpretation	How to present and interpret the data

8.1 Introduction

Close-inspection comprises a detailed visual and olfactory investigation of the PRF, for:

» Bats.

» Field-signs.

» Suitability.

8.2 Equipment

For a close-inspection to be comprehensive, it will in all but the most exceptional cases require:

» A ground-truthing account.

» A BTHK recording form.

» A camera.

» A torch.

» An endoscope.

» A tape-measure.

» A DBH tape.

In addition, a stock of sample-pots for storing collected droppings is helpful, along with some means of retrieving them, such as forceps, a long-handled tea-spoon or an HB-type Pooter.[1]

---

1 A suction device for retrieving insects that is operated by a hand-pressurised bulb in order that material does not enter the surveyor's mouth. The tool can be effective for retrieving dry bat-droppings with the minimum risk of contamination to the surveyor.

8.2.1 The ground-truthing account
This will have been created at the ground-truthing stage (see Chapter 6).

8.2.2 The BTHK recording form
The recording criteria adopted for the BTHK project was first defined by Sedgeley and O'Donnell (1999), then refined by Ruczyński and Bogdanowicz (2005), and finally expanded to include the substrate texture and cleanliness, odour and the presence of competing organisms.

An illustration of the BTHK recording form is provided at Table 8.1. An A4 size PDF copy can be downloaded from www.battreehabitatkey.com.

When submitting a record to the BTHK Database all the attributes should be assigned a value. Thus, when working through the texture, cleanliness and humidity sections, a tick or cross should be placed in *all* values.

8.2.3 The endoscope
If we accept that a professional survey should be sufficiently detailed to confidently detect all bat species, and all field-signs to establish whether it is 'more-likely-than-not' that a specific PRF is or is not a roost, the equipment should logically have the facility to inspect the full range of distances and record photographic representations that can be compared between visits and across the full surveillance period, and be made available for review by a third party.

Although bats may be found in situations where they can be seen without even the aid of a torch, this is relatively uncommon. The BTHK Database holds records of bats in positions above, to the side and below the entrance. The distances between the roost entrance and the bats in each of the three situations was as follows:

1. **Above the entrance** (sample size – 259 roosts, variously occupied by *Barbastella barbastellus, Myotis bechsteinii, M. daubentonii, M. nattereri, Nyctalus noctula, Pipistrellus nathusii, P. pipistrellus, P. pygmaeus, Plecotus auritus*, and *Rhinolophus hipposideros*):
» Mean average distance: 36 cm.
» Median distance: 22 cm.
» Range: 0–444 cm.

2. **To the side of the entrance** (sample size – 34 roosts, variously occupied by *Barbastella barbastellus, Myotis nattereri, Myotis* sp., *Nyctalus noctula, Pipistrellus pipistrellus, Pipistrellus pygmaeus* and *Plecotus auritus*):
» Mean average distance: 11 cm.
» Median distance: 8.5 cm.
» Range: 0–40 cm.

3. **Below the entrance** (sample size – 8 roosts, variously occupied by *Barbastella barbastellus, Myotis nattereri, Nyctalus noctula, Pipistrellus pipistrellus* and *Plecotus auritus*):
» Mean average distance: 16 cm.
» Median distance: 11 cm.
» Range: 0–44 cm.

The distance from the entrance is not the only factor; the shape of the internal chamber must also be taken into account. It is sufficiently uncommon for a PRF to open into a wide diameter with an even or tapering cylinder above and below, enabling a conclusive inspection to be achieved with a torch and a mirror. The more typical scenario is that there is a corner to negotiate or some other obstacle.

As a consequence, the greater proportion of surveillance will require an endoscope.

Table 8.1 The BTHK recording form

GROUNDSMAN*	**SITE NAME**					
	DATE					
	HABITAT (Phase 1 where known)					
	OS GRID REFERENCE					
	TAG NUMBER					
	TREE SPECIES					
	TREE ALIVE/DEAD					
	DBH (Diameter at Breast Height)					
	TREE HEIGHT					
	PRF on STEM/LIMB					
	PRF TYPE					
	DIRECTION PRF FACES					
CLIMBER	**MEASUREMENTS**	**PRF HEIGHT (m)**				
		DPH (Diameter at PRF Height)				
		ENTRANCE	**Height (cm)**			
			Width (cm)			
		INTERNAL	**Height (cm)**			
			Width (cm)			
			Depth (cm)			
	ROOSTING EVIDENCE	**Droppings: yes or no?** (look in base)				
		Bats (look in cavity)	**Species**			
			Number of bats			
			Torpid or awake			
			Above/below/to side of entrance			
			Distance from entrance (cm)			

INTERNAL CONDITIONS: (Tick as many of the following as are applicable. Where a field is negative please put a cross. There should be a value in EVERY box: ✓ or ✗)

	None	Pleasant	Not unpleasant	Unpleasant
SMELL				

	Smooth	Bobbly	Bumpy	Rough
SUBSTRATE				
	Clean** =	Waxy	Blackened	Polished
	Dirty** =	Dusty	Debris	Sludgy

	Dry	Damp	Wet	Mildew
HUMIDITY				

	Dome	Spire	Peak/wedge	Flat
APEX SHAPE				
	Chambered		Tube	

COMPETITORS (invertebrates, birds, mammals?)	

*	**PHOTOS**	1. Contextual photo showing tree in habitat	
		2. Close-up photo of the PRF	
	FORM	Check the form to ensure every box has a value	

* It is the Groundsman's responsibility to ensure the form is correctly and comprehensively completed.

** Cleanliness categories have a primary (i.e. clean or dirty) and then any associated secondary is ticked where it is observed.

The use of an endoscope is, however, intrusive and has the potential to be both damaging and disturbing, therefore:

» The inspection should only be undertaken where the surveyor is confident that there is no potential for damage to or destruction of the PRF.

» Where there is evidence to suggest bats may be present (such as droppings or an associated odour) the inspection should only be performed by a surveyor who is licensed[2] to disturb bats.

» Where there is no evidence to suggest bats may be present, but bats are encountered, if the surveyor is not licensed, he/she should withdraw immediately, while taking due care, and hand over the remaining aspect of the surveillance to a licensed surveyor (see more detailed discussion in Chapter 12).

There are various models of endoscope on the market, but the BTHK Project only recommends the Ridgid CA300 or above.[3] The reasons for this are as follows:

» The unit has the option of a 17-mm or 6-mm lens which can be changed in the field to suit the circumstances.

» The length of the camera 'snake' can be increased using 1-m and 3-m extensions.

» The camera has both zoom and image rotation.

» The camera can take still images and video footage which provide a photographic account of the inspection that can be compared with future inspections in order to search for changes in the internal environment. For example, if cobwebs and woodlice were present during the initial inspection, are gone at the next inspection, but present during the third, this hints that something larger than a woodlouse or a spider has been and gone between visits one and three.

» The CA300 has the best image quality of any portable unit we have used.

The CA300 does, however, have one significant flaw; the lack of an 'anti-glare' screen means that in bright sunlight it is often impossible to make out the image without the surveyor cloaking his/her head with a large loose hood or cowl. It is therefore recommended that a black pillow-case be modified and carried with the surveyor for exactly this purpose (no, this is not a joke).

8.3 Health and Safety

If the tree is to be ascended, it is up to the individual team to ensure they have the correct equipment, qualifications, insurance and risk assessment, and that their equipment is correctly maintained, regularly assessed and suitably certificated as safe to use.

From a health and safety perspective, it is wise to observe the PRF entrance for a short period prior to beginning the inspection to assess the potential that bees, wasps, hornets or tawny owls *Strix aluco*[4] might be present.

---

2   In order to allow data-gathering in support of Impact Assessments and for research, licences are made available from Natural England (NE), Natural Resources Wales (NRW), Department of Agriculture, Environment and Rural Affairs (DAERA) and Scottish Natural Heritage (SNH). To qualify for such a licence the surveyor must have demonstrated a robust understanding of bat ecology and competence in the methods covered by the licence.

3   All endoscopes used by the BTHK Project are purchased from Dart Systems (www.dartsystems. co.uk), who offer superb customer service.

4   Tawny owls nest in hollow trees in woodland, plantation, parkland, farmland and gardens (Holden and Cleeves 2002; Ferguson-Lees *et al.* 2011) and appear to favour the interior of decayed trunks (Thomson and Rankin 1923), but may also take shelter behind ivy (Sparks and Soper

From a legislative perspective, the potential for nesting lesser-spotted woodpeckers *Dryobates minor* to be present, and in the case of PRFs with larger entrances (in particular compression-forks) the potential for nesting barn owls *Tyto alba* should be considered.[5]

8.4 Close-inspection method

Close-inspections are not simply about searching for bats; close-inspections are about searching for all the clues that might hint that the PRF is exploited as a roost, and, in particular, anomalies – anything that is different which might be changes from one inspection to the next, or something that makes an individual PRF special. A thorough inspection will comprise five stages, as follows:

» **Stage 1** – Review.
» **Stage 2** – Search for droppings.
» **Stage 3** – Endoscope search.
» **Stage 4** – Odour test.
» **Stage 5** – Housekeeping.

By completing all five stages carefully, sufficient data will be provided for a meaningful interpretation.

8.4.1 Constraints

A full list of constraints was provided in the preceding chapter, but in summary the primary constraints associated with the efficacy of close-inspection are that:

» Not all PRFs can be comprehensively searched.
» Not all bats can be conclusively identified.
» There are few situations where a group of bats can be accurately counted.
» In many situations, the sex of the bats cannot be inferred from the numbers present.
» Where bats, larger competitors (i.e. grey squirrels) and any competitors that impose legislative restrictions upon the inspection (i.e. lesser-spotted woodpecker, barn owl, common dormouse) are present, it may be impossible to provide a comprehensive inspection of the PRF and thereby record all the values desired. Obviously, no attempt would be made to push past a roosting bat, but common sense should be exercised with respect to the potential for distress to any organism.

> **Note:** Any situation where any aspect of the inspection has proven inconclusive must be identified and careful consideration should be given to the potential merit of using an emergence method such as remote-observation and netting to gather missing data.

1970). A female tawny owl on a nest will attack repeatedly if disturbed and will not cease until the threat is well out of range. There is at least one recorded instance of an experienced ornithologist having been blinded in a tawny owl attack. If you're getting into the canopy you should also be aware that long-eared owls *Asio otus* nest within old squirrel *Sciurus* spp. dreys and display a comparable aggression to the tawny owl when defending young (Thomson and Rankin 1923).

5   Lesser-spotted woodpeckers and barn owls are listed on Schedule 1 of the *Wildlife & Countryside Act 1981 (& as amended)*. This legislation makes it a criminal offence to intentionally damage or destroy the nest or eggs of any wild bird, and intentionally *or recklessly* disturb any Schedule 1 bird while it is: building a nest; in, on or near a nest containing eggs or young; or to disturb dependent young.

Close-inspection: Stage 1 – Review

The first stage of every inspection should be the review of the results of the ground-truthing (see Table 6.3 in Chapter 6), and any pre-existing account of the PRF.

The ground-truthing result can be used to guide a search of the Database to identify field-signs and environmental cues associated with the bat species for which a *"reasonable likelihood"* of occurrence has been identified. This will ensure each inspection has an objective.

For example, if it is 'more-likely-than-not' the PRF might be exploited by Bechstein's bats, the objective questions might comprise the following:

» Are Bechstein's bats present?

» Is another bat species present that is known to 'time-share' roosts with Bechstein's bats?

» Are Bechstein's bat droppings present?

» Are bat-fly cases present?

» Are the roost entrance and internal dimensions sufficient for Bechstein's bats to gain access and is there sufficient room inside to accommodate a maternity group?

» Does the apex shape accord with Bechstein's bat roosts described on the Database?

» Does the environment offered accord with roosts occupied by Bechstein's bats?

» Is the PRF exploited by competitors that were also present in Bechstein's bat roosts at times when the bats were present or absent? Or, is another organism present that will preclude the presence of bats during a specific period, or even permanently?

As accounts of the inspections build, these might usefully be tabulated and the photographs of the internal environment collated into a format that can be taken into the field by the surveillance team. Thereafter, this review can be compared against the situation present on the current inspection to identify any changes.

It cannot be overemphasised: visual inspections should search for changes in the environment and occupancy by other organisms. If the accounts of previous inspections are not collated and reviewed by the survey team immediately prior to each subsequent investigation, the surveillance may lack a key element of focus. It is far better for a surveyor to be thinking *'if nothing has changed, I should be seeing a rough and dusty substrate in a dry environment'* and have a photograph of the historic situation, than to have no advance intelligence. Bats do not teleport into their roost positions but must crawl to their resting place. This means the substrate cues will be in evidence from the entrance and if the surveyor is simply searching for bats, they may miss environmental cues in their haste to get to the PRF base and apex. The statement *'no bats or droppings …'* gives little insight. The statement *'no bats, no droppings and no change in environment since the last inspection …'*, gives far greater confidence that nothing has been missed. Conversely, any change, no matter how minor, will immediately heighten attention and ensure the investigation is doubly thorough. *No one investigating a PRF does so in the hope that they won't find a bat roost!*

Close-inspection: Stage 2 – Search for droppings

Stage 2 should proceed as follows:

1. Perform a close search of the ground beneath the roost entrance using a cree torch.

2. Search any foliage in the field-layer beneath the roost entrance.

3. Search foliage and the upper surface of any branches in the shrub-layer.

4. Assuming the PRF isn't discontinuous and can therefore be scoped-out, search the PRF entrance and the base of the PRF, paying particular attention to any cobwebs present.

Figure 8.1 Individual droppings and accumulations in the context of each of the four stages of the search for droppings: ground, field-layer foliage, shrub-layer foliage and PRF entrance.

Figure 8.1 illustrates individual droppings and accumulations in the context of each of the four stages of the search for droppings.

Spider's-webs catch droppings that might otherwise be lost and appear to do an excellent job of preserving the DNA. Always investigate any spider's-webs very

Figure 8.2 Strata of alternate noctule and brown long-eared bat droppings.

thoroughly, and wherever possible take dropping samples from webs in preference to any other substrate.

Be aware: in contrast to houses, where droppings are made more conspicuous on flat, monotonal surfaces, and preserved in dry conditions away from coprophilous invertebrates, droppings in trees deteriorate more rapidly or are consumed, and tend to be camouflaged on an uneven, dropping-coloured substrate or covered by the droppings and frass from woodlice. If there is a lot of frass in the base of the PRF entrance, brush off the top layer to see if there is anything underneath. Frass is basically wood dust and does a good job of preserving droppings. In roosts that have been used for several years, there may be strata of droppings and it is worth having a dig about to prise the 'cork' of dry material out and then break it, as in the example in Figure 8.2.

In the winter, for some reason, old droppings may whiten, as in the examples at Figure 8.3, so look carefully and don't dismiss white droppings as avian.

Another rather strange autumn/winter occurrence is the fluffy mould cilia shown in Figure 8.4 that may appear on bat droppings but does not appear to form on bird droppings.

Figure 8.3 Whitened bat-droppings.

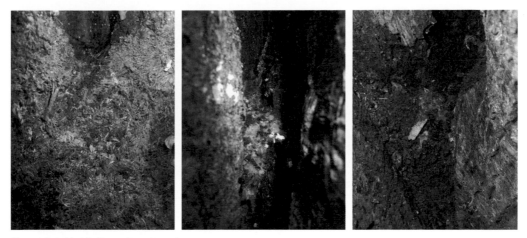

Figure 8.4 Mold cilia on bat-droppings.

Recording droppings

Any droppings encountered should be counted (if only an estimate) and a gauge of their age made as follows:

- » Black and shiny – Typically less than a fortnight.
- » Brown and dull – Typically over a fortnight old.

Where possible, a sample of droppings might be taken for DNA analysis but take note: European Bat Lyssaviruses (EBLVs) 1 and 2, commonly referred to as bat rabies, has been identified in Daubenton's bats in Surrey, Sussex and Lancashire, and in 2002 a bat worker in Angus, Scotland very sadly died from EBLV 2 infection. The virus is said to be spread through bites, scratches or through contact with mucous membrane, and web-hosted guidance suggests that if people do not handle bats they are not a risk (http://jncc.defra. gov.uk/page-3001-theme=textonly). Nevertheless, bat droppings are excrement and therefore a biohazard. As such, due care should be taken when retrieving droppings to test whether they are indeed bat droppings, and any subsequent storage and process for DNA analysis: **DO NOT TAKE RISKS**.

Collection in the field will typically comprise retrieving as much material as possible, but upon returning to the office it might be sensible to separate the material to ensure the freshest and most intact droppings are submitted for analysis, and frass, invertebrate remains, etc., are removed. This can be done under a dissection microscope but common sense and a good eye are typically adequate.

The most useful advice is to follow the instructions given by the various laboratories that offer the service.

Close-inspection: Stage 3 – Endoscope search

The endoscope search should be performed last, and should broadly follow this order:

1. Look down and search the base of the PRF.
2. Look on either side of the PRF entrance to check for concealed pockets or 'ram's-horns' (see Chapter 3 and in particular Figure 3.1). Any such features discovered should be carefully searched.
3. Look up.

Where bats are present, the BTHK project guidance is as follows:

1. Do not risk the endoscope coming into contact with a bat.
2. Withdraw immediately if babies or dependent young are present.
3. Where babies or dependent young are not present, the bat(s) should be in the light for no more than 10 seconds.
4. The inspection should be complete within 30 seconds.

No. 1 and No. 2 are **rules: do not *ever* risk the endoscope coming into contact with a bat and withdraw immediately if babies or dependent young are present.**

No. 3 and No. 4 are guidance and will depend upon circumstances. However, common sense should prevail and consideration given to how agitated any bat present is getting. The simple test of trial by social media can be helpful; ask yourself: *'Would I want anyone seeing how this inspection is going?'*

Keep in mind at this stage how detectable each bat species might be. For example, a maternity group of brown long-eared bats occupying a woodpecker-hole in July will leave a good deal more in the way of droppings than an individual noctule *Nyctalus noctula* occupying the same roost in a transitory capacity in April.

Barbastelles often hide in plain sight to one side of the entrance or in features that do not offer any more than partial concealment, and brown long-eared bats may rest on the back wall opposite the roost entrance, as demonstrated in the examples in Figure 8.5.

In addition, anecdotal accounts of roost inspections suggest that Daubenton's bats are particularly adept at concealing themselves in PRFs. Whereas most bats roost on the back wall above or below, but opposite the entrance, Daubenton's bats may also roost on the

Figure 8.5 Barbastelles (top row) and brown long-eared bats (bottom row) hiding in plain sight.

front wall immediately above the entrance, which requires the endoscope to be bent into a hook in order to see them. The test here is to ask yourself: '*If I have concluded the PRF is unoccupied, how nervous will I be that a bat might in fact be present if someone else inspects this immediately after me?*'

Recording

Bats

Where a bat or a group of bats is encountered, the orientation of bats with respect to the PRF entrance, and the distance the first bat is from the entrance should always be recorded. Obviously, where bats are present it may be impossible to record all the internal dimen-sions or the apex-shape, but if the PRF is vacant on the next inspection the missing value(s) should be collected. These values are useful in that they determine the maximum number of bats that might potentially be able to fit into the PRF, both for the species encountered, and any other species that might also exploit the PRF in another season (see below on 'time-share').

Where bats are present, an image should also be collected to establish whether con-clusive identification is possible using published keys (e.g. Dietz *et al.* 2011). Useful information to attempt to include in the photograph are the ear and tragus shape, the face shape and coloration, and the pelage, in particular, the dorsal and ventral fur coloration. With the common and soprano pipistrelles, the presence or absence of the internarial ridge is diagnos-tic so at least one 'nose-on' photograph in focus is helpful. If identification can be confidently demonstrated, then further action (i.e. remote-observation or static-netting) will be dictated by whether an accurate count can be achieved and the situation (which will also dictate whether sexing is required).

Unfortunately, the combination of the internal shape of many PRFs, the different ways in which groups of bats of different species congregate, the limitations of endoscope illumination, and the rule that the endoscope cannot be allowed to come into contact with a bat, mean that where groups of bats are present it is seldom possible to get any more than a threshold count (i.e. there are more than x bats present).

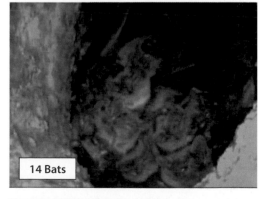

Figure 8.6 View of groups of bats in tree-roosts as seen on a Ridgid CA300 with a 17-mm lens, and the exact counts obtained from remote-observation or static-netting.

Where the endoscope view is from below, even in an open situation, only the front ranks of bats will be in view, but they will be moving: alternately putting their heads up to shout and then burying themselves behind a neighbour to hide from the light. Where the view is approaching from the side, the result is the same and in any case, where young are present, the adult females move to the front. If the PRF is a crevice, the typical count will be three bats, but there may be >20 behind that rank and there is no way of knowing this without using either remote-observation or static-netting. Figure 8.6 illustrates a typical view of bats in void and crevice situations with a count of exactly how many bats were actually present in each case obtained by remote-observation or static-netting.

Bat-fly cases

Bat-flies are non-flying parasites (Hutson 1984). The flies live on their hosts save for when the females have mature larva, at which point they leave the bat and deposit the larva on the internal substrate of the roost, sealing and gluing the larva to the roost wall in a sticky secretion that is soft and white when fresh but hardens to a red-brown colour (Hutson 1984). These puparia are shiny and their colour and appearance (either like tiny blobs of red-brown wax or blisters) make them conspicuous on the substrate when viewed with an endoscope (see Figure 8.7).

There are three bat-fly species that occur in Britain and each has a species-specific host association, as follows:

» *Basilia nana* is associated with Bechstein's bat.

» *Nycteribia kolenatii* is common in association with Daubenton's bat.

» *Phthiridium biarticulatum* occurs in association with brown long-eared bats and horseshoe-bats (Hutson 1984).

Unoccupied PRFs

Where bats, droppings or bat-fly cases are not present, interpretation will fall to consideration of:

» The entrance dimensions.

» The internal dimensions.

» The apex-shape.

» The internal environment.

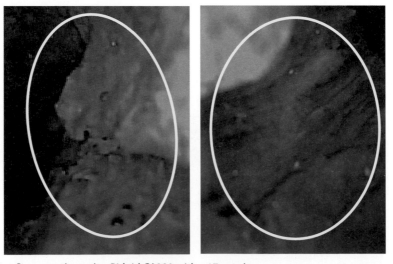

Figure 8.7 Bat-flies seen through a Ridgid CA300 with a 17-mm lens.

Entrance dimensions and DPH

The height and width of the entrance are recorded as shown at Figure 8.8, using a tape-measure to the nearest 0.5 cm, unless the entrance is below 2 cm on any plane, in which case the entrance is recorded to the nearest millimetre.

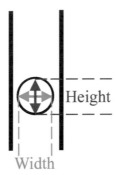

Figure 8.8 Recording the entrance dimensions. If you're in doubt about what constitutes the entrance, try thinking of it in terms of where you might most cost-effectively fit a door to secure the PRF against a burglar.

The Diameter at PRF Height (DPH), coupled with the vitality (i.e. whether the tree/ limb is alive or dead) and the internal width of the PRF give an insight into its structural strength, probable longevity and a gauge of whether repeat inspections might be hazardous. This is particularly pertinent when considering a recommendation for a pre-felling inspection that might be a significant period in the future. The DPH is measured using the diameter tape across the centre of the entrance.

Internal dimensions

The internal dimensions are recorded using the endoscope as a guide, and then compared with measurements taken with an open tape-measure as illustrated at Figure 8.9.

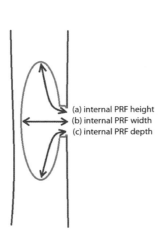
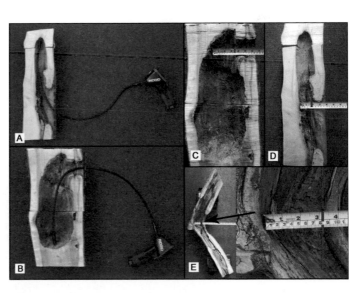

Figure 8.9 Recording the internal dimensions. Left: all measurements are taken from the bark side of the entrance. Right: (A) internal height; (B) internal depth; (C) woodpecker-hole internal width; (D) tear-out internal width; and (E) hazard-beam internal width.

Apex-shape

Where the PRF extends above the entrance, the apex-shape may influence the species that may be present and also the reason the PRF is occupied (i.e. winter, maternity, mating, etc.). The Database sorts apex-shapes into four enclosed types and one open type comprising:

» Closed-type:

 – Dome;

 – Spire;

 – Peak/wedge; and

 – Flat (i.e. there is no upward development above the entrance, or the PRF culmi- nates in a clearly flat horizontal substrate).

» Open-type:

 – Tube.

An additional consideration is whether the apex culminates in an individual apex or more than one. The later situation is identified as 'chambered' and this category will always encompass one or more of the five apex shapes; for example, there might be two individual spires in which bats might roost, or a spire and a dome, etc. Figure 8.10 illus- trates the different apex shapes.

Internal environment

Analysis of records held on the Database indicates that certain bat species have associa- tions with particular environmental conditions in different periods of the year.

The inspection of the internal substrate should assess whether it deviates from that expected. However, this may not be obvious, even if the PRF is a roost but has not been occupied for several months. During the time that bats are absent, the PRF will be occupied by other organisms, and those organisms make a good deal of mess. In order to assess the substrate it is often necessary to push through the mess of spider's-webs, snail-shells, bird's-nests and general detritus to see the potential beneath, and ask the question: *'Is this mess obscuring an environment that is associated with roosting bats?'*

To illustrate, the images on the left in Figure 8.11 are of a roost that is exploited by both Natterer's and brown long-eared bats. From October to November brown long-eared bats were present, but in the following January the bats were gone and the roost was entirely clogged with garden snails and a dead blue tit. The same situation is evident in the centre images; noctules are present in December, but by August the roost is full of webs. Reversing the situation, the images on the right are of a mucky situation in October, but suddenly clean in the following January and a wintering noctule is present in the apex. All these roosts are occupied by the same species in the same periods year-on-year.

The BTHK Database divides environmental cues into five categories:

» Humidity.

» Texture.

» Cleanliness.

» Competitor occupancy.

» Odour.

Humidity

Humidity is recorded as one of three characteristics:

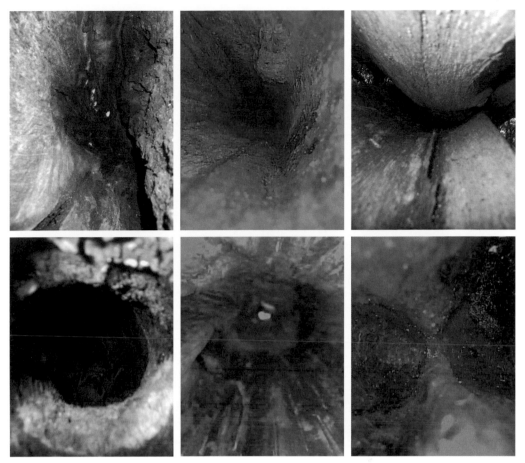

Figure 8.10 The six different apex shapes: Top left: domed noctule roost. Top middle: spired Natterer's bat roost. Top right: peaked brown long-eared bat roost. Bottom left: flat-topped noctule roost. Bottom middle: tubular Bechstein's bat roost. Bottom right: chambered-spire exploited by individual Natterer's and brown long-eared bats.

1. **Dry** – Characterised by an arid substrate with no question of any moisture (see Figure 8.12).

2. **Damp** – Characterised by a humid environment (see Figure 8.13).

3. **Wet** – Characterised by water-droplets and surface-flow on walls and sometimes even liquid pooled in the base (see Figure 8.14).

Texture

Texture comprises four subcategories:

1. **Smooth** – Any situation where the internal substrate is smoothed flat so that it has no sharp or jagged edges and no coarse-grained projections (see Figure 8.15).

2. **Bobbly** – Any situation where the substrate is finely stippled like the tiny speckling of embossed braille (see Figure 8.16).

3. **Bumpy** – Any situation where the substrate is uneven but not coarse-grained. Bumpiness ranges from marbling (much like the wet surface of a limestone cave) through rounded projections (see Figure 8.17).

4. **Rough** – Any situation where the substrate is course-grained and jagged (see Figure 8.18).

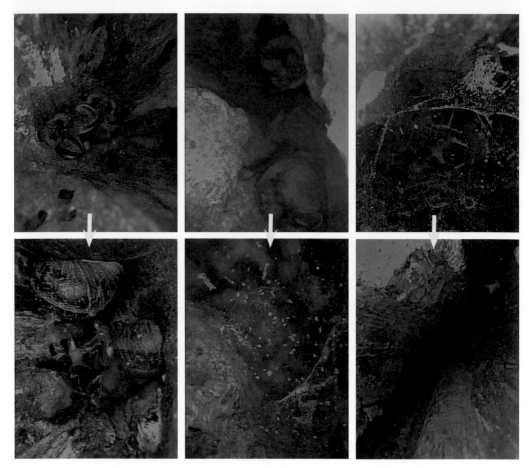

Figure 8.11 Left: a roost which held brown long-eared bats in November, garden snails and a dead blue tit *Cyanistes caeruleus* in January, but a Natterer's bat in the following May. Middle: noctules present in December and the roost again clogged with webs in the following August. Right: a reversal of the situation – the roost is clogged with webs in October but clean when occupied by a noctule in the following January.

Cleanliness

Cleanliness is recorded as one of eight characteristics, comprising:

1. **Clean** – Cleanliness is associated with dry, damp and wet surfaces, and characterised by a lack of cobwebs, loose dust/debris or any accumulation of humus or fungal hyphae or bacterial exudation (see Figure 8.19).

2. **Waxy** – Waxiness is associated only with dry surfaces, and characterised by a greasy tallow on a dry substrate resulting in a sheen that may be variously white, grey and even ginger/orange (see Figure 8.20).

3. **Blackened** – Blackening is associated only with dry surfaces, and characterised by an oily blackening that highlights the striations in dry ray-wood. Note: all damp and wet PRFs appear blackened but this is an artefact of the environment and does not qualify as this category (see Figure 8.21).

4. **Polished** – Polishing is associated only with dry surfaces, and characterised by a glossy appearance to the wood that may be so pronounced as to appear lacquered (see Figure 8.22).

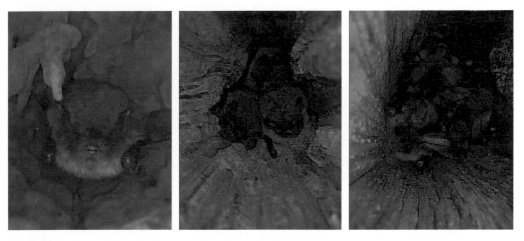

Figure 8.12 Examples of dry PRFs. Left: Daubenton's bat. Middle: common pipistrelles. Right: brown long-eared bats.

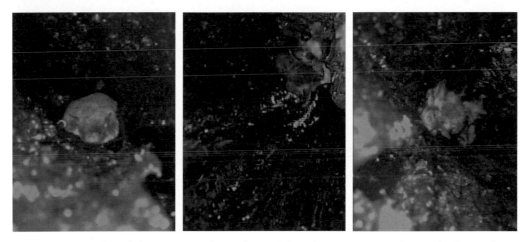

Figure 8.13 Examples of damp PRFs. Left: Daubenton's bat. Centre: Natterer's bat. Right: brown long-eared bats.

5. **Dirty** – Dirtiness is associated with at least historically or seasonally damp surfaces, and characterised by small particles of dry to damp material stuck to the internal substrate, such as may rub-off on the endoscope, requiring it to be withdrawn and cleaned (see Figure 8.23). Dirt of this nature would typically be disturbed/rubbed-off by any animal passing across it.

6. **Dusty** – Dustiness is associated only with dry surfaces, and characterised by a surface layer of small particles of sawdust smaller than chips (see Figure 8.24).

7. **Debris** – Debris is associated only with dry surfaces, and characterised by accumulations of larger loose detritus, frass and wood chips, which may clog the internal void and have to be pushed aside (see Figure 8.25).

8. **Sludgy** – Sludginess is associated only with damp and wet PRF, and characterised by bacterial exudations of thick mucus-like semiliquid humus, often with a pool of slime in the base of the PRF (see Figure 8.26).

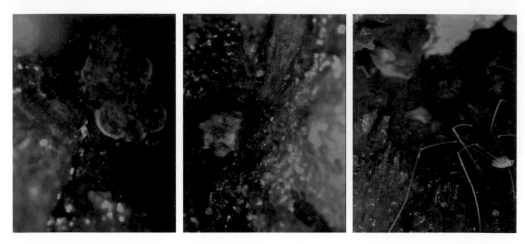

Figure 8.14 Examples of wet PRFs with bats lying to one side of seeping water. Left: Daubenton's bats. Centre: brown long-eared bat. Right: brown long-eared bat.

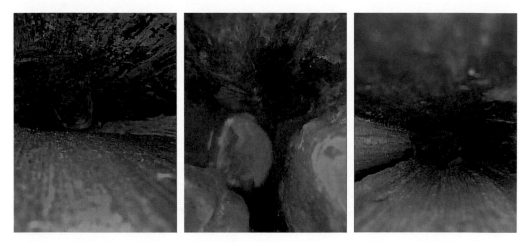

Figure 8.15 Examples of smooth substrates in tree-roosts.

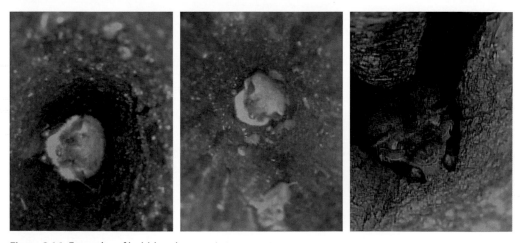

Figure 8.16 Examples of bobbly substrates in tree-roosts.

Figure 8.17 Examples of bumpy substrates in tree-roosts.

Figure 8.18 Examples of rough substrates in tree-roosts.

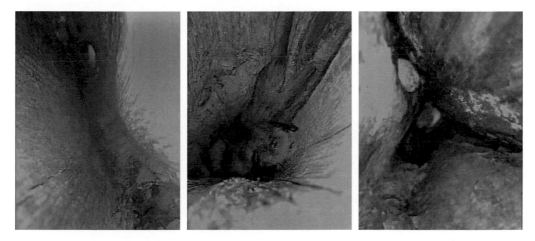

Figure 8.19 Examples of clean substrates in PRFs.

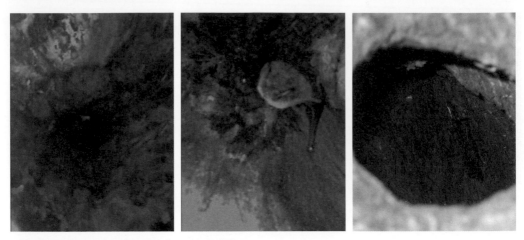

Figure 8.20 Examples of waxy substrates in PRFs.

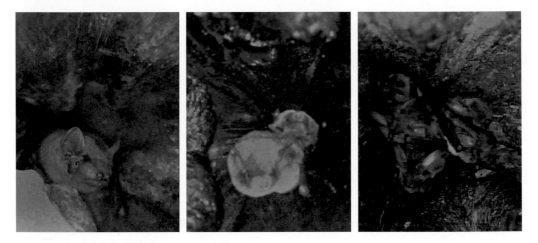

Figure 8.21 Examples of blackened substrates in PRFs.

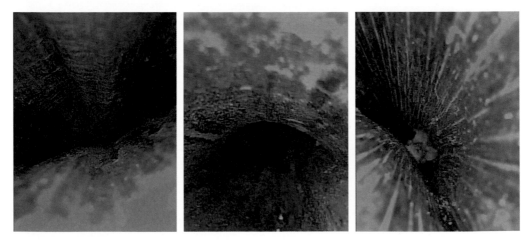

Figure 8.22 Examples of polished substrates in PRFs.

Figure 8.23 Examples of dirty substrates in PRFs.

Figure 8.24 Examples of dusty substrates in PRFs.

Competitor occupancy

The presence of some competitors, such as woodlice, lipped snails and blue tits *Cyanistes caeruleus*, may be an indication that the environment the PRF offers is favourable for roosting bats.

Turning this on its head, the absence of invertebrates, particularly woodlice, spiders and slugs, which are present in most PRFs, should prompt the question as to what cleared them out, even if no bats are present that day. This is particularly relevant to repeat inspections: if invertebrates were present in abundance on the first inspection, and they are gone on the second, why have they gone? Is it just a seasonal artefact (such as an invertebrate overwintering as larvae and therefore being functionally undetectable) or has something displaced them?

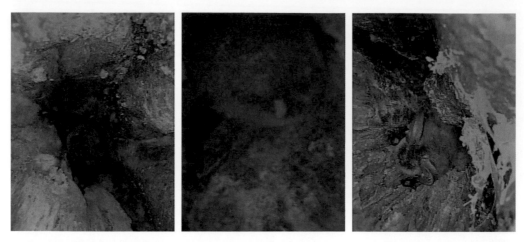

Figure 8.25 Examples of debris in PRFs.

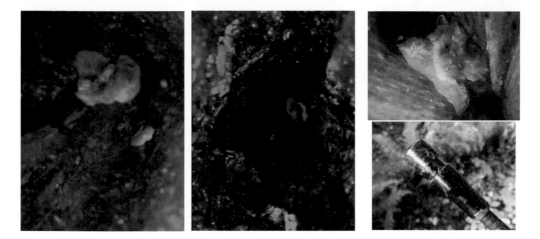

Figure 8.26 Examples of sludgy substrates in PRFs.

Competing organisms in trees comprise:

» **Invertebrates** – Woodlice, slugs and snails, flies, median wasps *Dolichovespula media*, hornets *Vespula crabro*, tree bumblebees *Bombus pratorum*, hive bees *Apis mellifera*, spiders, millipedes, centipedes, moths, butterflies and beetles.

» **Amphibians** – Common toad *Bufo bufo* and smooth newt *Lissotriton vulgaris*.

» **Birds** – Most frequently: blue tits, great tit *Parus major*, pied flycatcher *Ficedula hypoleuca*, wren *Troglodytes troglodytes* and jackdaws *Corvus monedula*.

» **Mammals** – Wood mice *Apodemus sylvaticus*, hazel dormice *Muscardinus avellanarius* and grey squirrels *Sciurus carolinensis*.

Figure 8.27 illustrates some of the invertebrate competitors that are frequently encountered in PRFs, and Figure 8.28 illustrates some of the larger competitors that may be encountered in PRFs.

Figure 8.27 Some of the invertebrate competitors that are frequently encountered in PRFs. Top line, left to right: woodlice *Porcellio scaber*, tree slugs *Lehmannia marginata* and dark-lipped snails *Cepaea nemoralis*. Middle line, left to right: hornets *Vespa crabro*, median wasp *Dolichovespula media* and honey bee *Apis mellifera*. Bottom line, left to right: white-legged snake millipede *Tachypodoiulus niger*, funnel-web spider *Amaurobius* sp. with common pipistrelles, and a brown long-eared bat above a spider's-web.

Figure 8.28 Some of the larger competitors that may be encountered in PRFs. Top line, left to right: common toad *Bufo bufo* (up to 3 m from ground-level), green woodpecker *Picus viridis* and nesting great tit *Parus major* (top) and pied flycatcher *Ficedula hypoleuca* (bottom). Middle line, left to right: blue tit *Cyanistes caeruleus* (typical view of night-roosting perch above the entrance), blue tit chicks and wren *Troglodytes troglodytes*. Bottom line, left to right: wood mouse *Apodemus sylvaticus*, hazel dormouse *Muscardinus avellanarius* and grey squirrel *Sciurus carolinensis* (with brown long-eared bat).

8.4.2 Close-inspection: Stage 4 – Odour

Odour may indicate the current or historic presence of bats, even in the absence of droppings.

Having established the PRF does not hold any organism that might attack the surveyor, Stage 3 of the close-inspection comprises an 'odour test' to search for any unexpected anomaly. This is performed by the surveyor getting his/her nose into the PRF entrance and taking a series of shallow sniffs.

Initially it helps to ask whether the odour is what the surveyor was expecting. Expected odours typically comprise the smell of wood, moss, leaf-litter or no discernible odour at all (the latter being more common in winter). Unexpected odours encompass anything that strikes the surveyor as odd, and can be broadly subdivided into the following categories:

» **Pleasant** – An odour that the surveyor finds entirely acceptable and would seek exposure to (for example, pleasant odours that have been reported to the BTHK Database include: citrus, sweet, hay, mushrooms, sweet and sour sauce, and Worcestershire sauce).

» **Not unpleasant** – An odour that the surveyor finds tolerable but would not seek exposure to (for example, not unpleasant odours reported to the BTHK Database include: peanut butter, wet tar, burnt rubber, damp cellar, dry attic, musk, musty, smoky, rodents (anyone who has kept a hamster or gerbil will be familiar with this odour), earthy, tanalised timber and tar).

» **Unpleasant** – An odour that is intolerable to the point of being repellent (for example, unpleasant odours reported to the BTHK Database include: sickly sweet and sour, funky, burnt feathers, grass snakes, fish and urine).

Table 8.2 Odour descriptions recorded in accounts of roost inspections held on the BTHK Database (bat species not listed have either no data on the Database or insufficient samples to be able to provide a reliable description)

SPECIES	EXPECTED/ UNEXPECTED		ODOUR TYPE(S)			DESCRIPTION
	Expected	Unexpected	Pleasant	Not-unpleasant	Unpleasant	
Barbastelle	—	✓	✓	✓	—	Sweet or musky
Daubenton's bat	—	✓	—	—	✓	Foul and grass snakes
Natterer's bat	—	✓	✓	✓	—	Freshly broken mushrooms through to wet tar and acrid burnt rubber
Noctule	—	✓	—	✓	✓	Wet tar and burnt rubber through to grass snakes
Common pipistrelle	—	✓	—	✓	✓	Funky, musty and sickly sweet and sour
Brown long-eared bat	—	✓	✓	✓	✓	Sweet, through to spicy, smoky, burnt-rubber and even a similar smell to caged gerbils

More than any other aspect of the inspection, the odour-test highlights the fact that inspections are not just about searching for bats; they are about searching for anomalies between PRFs and between visits. Even in the absence of the bats and their droppings, some bat species leave behind an odour that is conspicuous in comparison to the odours present in unoccupied PRFs, and is also distinctive between bat species and also between competitor species. Table 8.2 sets out odour descriptions recorded in accounts of roost inspections held on the BTHK Database.

Recording odour

For the odour-test to have merit the result should be stratified as follows:

» Step 1:
  – Expected; and

  – Unexpected.

» Step 2:
  – No discernible odour;

  – A pleasant odour;

  – An odour that is not unpleasant; and

  – An unpleasant odour.

» Step 3:
  – Describe what the odour is like using any other examples of the odour that are familiar to the surveyor. It is also helpful to make a note of how strong the odour is. Olfactory memories are particularly strong and particularly resistant to interference. As a result, a good description will be reliable between visits, providing the same surveyor surveys the same tree each time and reviews the records of the previous inspection to search for the same odour.

8.4.3 Close-inspection: Stage 5 – Housekeeping

When the inspection is complete, if it is possible, frass and old droppings might be swept from the entrance so that any new evidence of occupation is immediately evident and the surveyor will not be left wondering whether the evidence was left over from the previous inspection.

8.5 Interpretation

8.5.1 General

Interpretation comprises the collation of the results in a format that is more helpful for practical application, followed by a comparison of the individual results in respect of each recording criterion against associations identified within records held on the Database.

8.5.2 Presentation of inspection results

The objective questions at this stage might comprise:

1. Were bats encountered allowing a conclusive account of their species, sex and numbers?

2. Were droppings or bat-flies that give conclusive evidence that the feature is exploited by roosting bats, but may not have provided a conclusive account of all three data (i.e. the species, sex and numbers) recorded?

3. Does the PRF comprise a combination of characteristics which support the hypothesis that it is 'more-likely-than-not' the PRF is suitable for the bat species identified at the desk-study stage?

4. Did the results of the individual inspections demonstrate changes, or were they constant?

Objectives 1, 2 and 3 are easily identified, but objective 4 is less so, and the interpretation of environmental data benefits from visual presentation.

The first stage of the interpretation might therefore be to present the data in such a way as to make any anomaly that might have a positive bearing conspicuous. As the interpretation is progressive, it might follow the format of Table 8.3 (a Word copy of this table can be downloaded from www.battreehabitatkey.com), which makes any environmental change from one visit to the next immediately obvious. This is of material benefit both during the surveillance and at the final interpretation stage (see Stage 1 – Review).

8.5.3 Interpretation steps

General

In broad terms, the interpretation can be divided into two avenues:

» Positive encounters with bats, droppings and/or bat-fly cases.

» Negative outcomes which may be compared with the accounts provided to infer the suitability of the PRF to individual bats in each season, within the margin of *"reasonable likelihood"*.

The data that are considered in both situations are as follows:

POSITIVE ENCOUNTERS	NEGATIVE OUTCOMES	
	Dimensions	**Environment**
Bats	Entrance dimensions	Humidity
Droppings	Internal dimensions	Texture
Bat-fly cases	Apex shape	Cleanliness
		Competitors
		Odour

Table 8.3 A template for the presentation of repeat close-inspection data

Tree reference no.:								
CATEGORY	**CRITERIA**	**DATE**						
BATS	Bats species							
	Bat numbers							
	Above entrance: a Below: b Side: s							
	Distance from entrance (cm)							
	Awake or torpid							
DROPPINGS PRESENT								
BAT-FLIES PRESENT								
ENTRANCE DIMENSIONS	Height							
	Width							
INTERNAL DIMENSIONS	Height							
	Width (i.e. diameter)							
	Depth							
INTERNAL VOLUME*								
APEX SHAPE								
HUMIDITY	Dry							
	Damp							
	Wet							
ODOUR	None							
	Pleasant							
	Not unpleasant							
	Unpleasant							
TEXTURE	Smooth							
	Bobbly							
	Bumpy							
	Rough							
CLEANLINESS	Clean							
	Waxy							
	Blackened							
	Polished							
	Dirty							
	Dusty							
	Debris							
	Sludgy							
COMPETITORS PRESENT								

*Simply use the Google online calculator; type in 'calculating the volume of a cylinder' (if it is a dome apex), a cone (if it is a spire apex), a wedge (if it is a peak apex) or a rectangle (if it is a crevice-type PRF with a flat apex).

8.5.4 Positive encounters

General

Even positive encounters with bats may have a bearing beyond the species present.

In the case of droppings, additional DNA analysis may be required. In the case of bat-fly cases, caution must be applied when attempting to assign the flies to a specific host. In both cases significant constraints must be acknowledged; in particular, the fact that the absence of positive encounters with either droppings or bat-fly cases cannot be used to infer roost status in the negative.

Bats

Where bats are encountered and the species can be identified with certainty, the results may be extrapolated to give a wider hypothesis of other species that might also exploit the feature. This is because an individual PRF may be occupied by different bat species for different purposes at different times of year, and these associations are identifiable in the records on the Database. Considering the PRF type, internal dimensions, characteristics and apex shape, these associations can be used as a predictive tool. Table 8.4 shows which species are known to 'time-share' PRF with other species. However, it is recommended that reference is made to the most up-to-date Database available to identify when each species was present, in what numbers and for what purpose.

Table 8.4 Bat species recorded exploiting the same roost at different times of year

	B. ba	M. be	M. al	M. br	M. da	M. na	M. my	N. le	N. no	P. na	P. pi	P. py	P. au	R. hi
B. ba						✓			✓	✓	✓	✓		
M. be					✓	✓			✓				✓	
M. al														
M. br										✓	✓	✓	✓	
M. da		✓				✓	✓	✓	✓		✓	✓	✓	
M. na	✓	✓			✓				✓		✓	✓	✓	
M. my					✓									
N. le					✓				✓					
N. no	✓	✓			✓	✓		✓					✓	
P. na	✓			✓								✓		
P. pi	✓			✓	✓	✓								
P. py	✓			✓	✓	✓					✓			
P. au		✓		✓	✓	✓			✓					
R. hi														

Droppings

There are significant constraints in the interpretation of negative results. This is because in 78.7% (666 of 846) of encounters of bats in trees in records held on the BTHK Database no droppings were recorded. The reality is that in damp and wet roosts droppings will simply turn to sludge and be undetectable, and in many other situations they simply drop to the woodland floor and are undetectable. This means that droppings can only be used as a positive, and the lack of droppings cannot be used to infer roost absence.

The discovery of droppings outside a roost requires careful interpretation. Experience has shown that applying a 'scepticality test' and attempting to identify whether there is any other explanation for the situation the dropping(s) has been found in is helpful. For example, might a bat momentarily resting during a bout of foraging have deposited the dropping, or a bat flying over, or a bat exiting a roost in an adjacent building or tree?

Where droppings are discovered *inside* the PRF (even in the absence of bats) it can reasonably be concluded that the PRF is a day-roost (Andrews and Gardener 2016). However, the 'scepticality test' is again useful when there is a temptation to assign a roost status, i.e. night- or day-roost, individual bat or aggregation.

Colonies of several bat species that roost in both houses and trees differ in the size of their groups in the two situations. For example, the entirety of a nursery colony of female brown long-eared bats may occupy a single roof void in a building, but that same colony may be split into three sub-groups when occupying trees, with a resulting drop in numbers in any one roost. Furthermore, the occupancy of roosts in houses appears to be more constant than it is in trees (i.e. although individual bats may come and go from day to day, and numbers may fluctuate widely, it is more common for a roost in a house to be constantly occupied by the same species over a longer duration than is the case with a roost in a tree). There may therefore be significantly greater numbers of droppings in roof-voids due to the higher number of bats present, the constant occupancy and the overall longer duration of occupancy than may be typical for trees.

It is therefore wise to limit interpretation to simply identifying that droppings are indeed bat-droppings, and using any good-quality material to seek DNA identification of species. This leads to the final 'scepticality test': any temptation to attempt to identify droppings without DNA analysis should be avoided. Any such record submitted to the BTHK Database requires an accompanying DNA result. However, if the DNA result is inconclusive the record may be accepted as 'unidentified bat species' if it is accompanied by a macro photograph that allows the image to be enlarged to such an extent that there is at least one scat in focus that is clearly a bat dropping. This is because not all droppings that look like bat droppings may in fact be bat droppings. Regardless, without conclusive identification to species (and in the absence of any other diagnostic field-sign) any attempt to interpret the result may be speculative in the extreme, and therefore open to challenge.

Table 8.5 illustrates the percentage of inspections where bats were present that also encountered droppings, in terms of the numbers of bats set out as individuals, 2–4 bats and 5 or more for each of the seasons.

Table 8.5 The percentage of inspections where bats were present that also encountered droppings in terms of the numbers of bats set out as individuals, 2–4 bats and 5 or more for each of the seasons

NUMBER OF BATS	SEASON					
	Winter (Jan/Feb)	Transitory (spring-flux Mar/Apr)	Maternity		Mating (Sep/Oct)	Transitory (autumn-flux Nov/Dec)
			Pregnancy (May/Jun)	Nursery (Jul/Aug)		
1	2.9	8.6	9.6	12.2	14.3	5
2–4	6.7	18.8	50	21.7	20	18.8
5+	0	23.1	50	33.9	41.2	30

Bat-fly cases

The presence of bat-fly cases is conclusive evidence that the PRF is exploited as a roost. However, caution must be exercised when attempting to assign the associated species. In some cases, the geographical situation may narrow the possibilities due to the restricted distribution of Bechstein's bat. In others, the dimensions of the entrance may preclude the presence of horseshoe-bats. However, the overlap in habitat and PRF exploitation exhibited by Bechstein's bat, Daubenton's bat and the brown long-eared bat (which have even been recorded using the same roost-feature in the same tree on different days) should be acknowledged as a constraint that may confound analysis to species level, and will mean that consideration of additional cues (such as odour and humidity) will be required in support of any further hypothesis.

8.5.5 Negative outcomes

General

Where conclusive evidence is not recorded, the data may be used to assess the suitability of the PRF for individual bat species in each of the six seasons. In addition, a rough (but nevertheless reasoned) hypothesis can also be defined as to the number of bats the PRF might be suitable to hold. This is achieved by comparing the data against the individual species accounts provided, which were defined using the records on the BTHK Database.

   The comparison accounts are divided into dimensions and environmental characteristics.

Dimensions

In the dimension category, the comparative data are given as numerical ranges as follows:

- » The full range of the recorded parameters.
- » Defining the interquartile range (the middle 50%) of recorded values thereby providing thresholds in a focus range of above 50% to give a *"reasonable likelihood"* interval.
- » Identification of the median as a middle value.

Environment

In the environment category, the comparative data are given as a percentage of applicable records that fell into the category as well as the absolute number of records held on the BTHK Database (in parentheses). Where more than 50% of the records accord with an attribute, this is identified as *"reasonable likelihood"* with the figure presented in bold white on a black background. It should be noted that an individual PRF may encompass more than one attribute in any category (e.g. many PRFs that are dry near the entrance are damp in the apex, and many have sections that are smooth and others that are rough).

8.5.6 Acceptance of omissions and shortfalls in the data

Omissions

At the time of writing the Database holds no records of Alcathoe's bat or whiskered bat, and is data deficient in respect of Brandt's bat and Leisler's bat. Accounts for these species have therefore been omitted.

Shortfalls in data

It is accepted that the *"reasonable likelihood"* (i.e. 'more-likely-than-not') threshold is somewhat 'rustic' as a predictive tool, and it will not be infallible. It is also accepted that it is incomplete because the Database holds no records in respect of Alcathoe's bat and whiskered bat, and is data deficient in respect of Brandt's bat and Leisler's bat, and also because there are only low numbers of records of some other species, and for maternity and all-male groups.

Notwithstanding, the framework is comprehensively evidence-supported and in many situations, there is sufficient data to identify which characteristics and conditions suggest it is 'more-likely-than-not' a specific PRF is suitable for a specific bat species, and when they may exploit it. Furthermore, the accounts identify a range of competitors that exploit the same conditions as the individual bat species.

Table 8.6 Barbastelle *Barbastella barbastellus*: associated dimensions (cm)

ENTRANCE HEIGHT	Winter	Spring-flux	Pregnancy	Nursery	Mating	Autumn-flux
Total range	185–185	3–185	4.8–16	4.8–16	30–87	9–140
Reasonable likelihood	185–185	20.1–106.7	7.2–10	7.2–10	35–63.5	10–65
Median	185	67.5	8	8	40	23.5
ENTRANCE WIDTH	Winter	Spring-flux	Pregnancy	Nursery	Mating	Autumn-flux
Total range	1.5–1.5	1.2–9.5	2–5	2–5	2–4	0.8–3.5
Reasonable likelihood	1.5–1.5	1.5–3.8	2–3.3	2–3.3	2.7–3.7	1.5–2
Median	1.5	2	2.4	2.4	3.5	1.5
INTERNAL HEIGHT	Winter	Spring-flux	Pregnancy	Nursery	Mating	Autumn-flux
Total range	9–9	0–58	0–58	0–58	2–80	0–58
Reasonable likelihood	9–9	12–28.7	30–46	30–46	21–60	11.5–20.5
Median	9	20.5	41	41	40	13
INTERNAL WIDTH	Winter	Spring-flux	Pregnancy	Nursery	Mating	Autumn-flux
Total range	18.2–18.2	3–18.2	2.1–7	2.1–7	3–5	0–10
Reasonable likelihood	18.2–18.2	4.5–10.5	2.1–7	2.1–7	3.5–4.5	2–6
Median	18.2	7	6	4.5	4	3
INTERNAL DEPTH	Winter	Spring-flux	Pregnancy	Nursery	Mating	Autumn-flux
Total range	0–0	0–100	69–101	69–101	0–26	0–3.5
Reasonable likelihood	0–0	0–0	81–101	81–101	0–13	0–0
Median	0	0	93	93	0	0
INTERNAL VOLUME	AVERAGE LENGTH OF BAT (top of head to rump) (mm)	AVERAGE WIDTH OF BAT (wrist to wrist) (mm)	AVERAGE CONDYLO-BASAL OF BAT (nose to back of skull) taken from Dietz *et al.* 2011 (mm)	AVERAGE VOLUME OF BAT (cm³)	TREE ROOST MATERNITY COLONY SIZE	MINIMUM VOLUME REQUIRED FOR A MATERNITY COLONY (cm³)
	67.5	34	19.4	44.5	22–54 (BTHK Database)	979–2,403
APEX SHAPE	Winter	Spring-flux	Pregnancy	Nursery	Mating	Autumn-flux
Dome	—	—	—	—	33.3 (1)	—
Spire	—	25 (2)	—	—	33.3 (1)	40 (2)
Peak	100 (1)	75 (6)	—	—	33.3 (1)	40 (2)
Flat	—	—	—	—	—	20 (1)
Chambered	—	12.5 (1)	—	—	—	20 (1)
Tube	—	—	—	100 (2)	—	—

Table 8.7 Barbastelle *Barbastella barbastellus*: associated environment

HUMIDITY	Winter	Spring-flux	Pregnancy	Nursery	Mating	Autumn-flux
Dry	100 (1)	100 (7)	100 (1)	100 (2)	100 (2)	100 (5)
Damp	—	—	—	—	—	—
Wet	—	—	—	—	—	—
TEXTURE	**Winter**	**Spring-flux**	**Pregnancy**	**Nursery**	**Mating**	**Autumn-flux**
Smooth	100 (1)	71.4 (5)	100 (1)	100 (2)	-	40 (2)
Bobbly	—	—	—	—	—	—
Bumpy	100 (1)	42.9 (3)	—	—	—	20 (1)
Rough	—	57.1 (4)	100 (1)	—	100 (2)	100 (5)
CLEANLINESS	**Winter**	**Spring-flux**	**Pregnancy**	**Nursery**	**Mating**	**Autumn-flux**
Clean	100 (1)	85.7 (6)	100 (1)	100 (2)	—	80 (4)
Waxy	100 (1)	71.4 (5)	100 (1)	100 (2)	—	40 (2)
Blackened	—	57.1 (4)	—	—	—	40 (2)
Polished	—	14.3 (1)	—	—	—	—
Dirty	—	—	—	—	50 (1)	—
Dusty	—	28.6 (2)	—	—	—	60 (3)
Debris	—	14.3 (1)	—	—	—	20 (1)
Sludgy	—	—	—	—	—	—
ASSOCIATED COMPETITORS (i.e. species that may time-share PRF with barbastelles)	Unidentified spiders and spider's-webs					
ODOUR	**Winter**	**Spring-flux**	**Pregnancy**	**Nursery**	**Mating**	**Autumn-flux**
None	100 (1)	100 (6)	100 (1)	—	50 (1)	80 (4)
Pleasant	—	—	—	100 (2)	—	—
Not unpleasant	—	—	—	—	50 (1)	20 (1)
Unpleasant	—	—	—	—	—	—

Table 8.8 Bechstein's bat *Myotis bechsteinii*: associated dimensions (cm)

ENTRANCE HEIGHT	Winter	Spring-flux	Pregnancy	Nursery	Mating	Autumn-flux
Total range	—	6–90	6–76	3–230	6–71	—
Reasonable likelihood	—	27–69	8–9.5	7–42.2	7.7–13	—
Median	—	48	8	15.6	9.2	—
ENTRANCE WIDTH	Winter	Spring-flux	Pregnancy	Nursery	Mating	Autumn-flux
Total range	—	3.5–6	1.3–24	2–13	2–9.5	—
Reasonable likelihood	—	4.1–5.3	5–7	4.5–6.3	3.6–6.1	—
Median	—	4.7	6.5	5.5	5	—
INTERNAL HEIGHT	Winter	Spring-flux	Pregnancy	Nursery	Mating	Autumn-flux
Total range	—	30–90	18–147	0–230	4–90	—
Reasonable likelihood	—	45–75	37–90	12.5–44	42–68.2	—
Median	—	60	80	27	51	—
INTERNAL WIDTH	Winter	Spring-flux	Pregnancy	Nursery	Mating	Autumn-flux
Total range	—	5–12	6–20	3.8–28	4–40	—
Reasonable likelihood	—	6.7–10.2	8–18	7.2–16.1	7–16	—
Median	—	8.5	14.5	9.7	7	—
INTERNAL DEPTH	Winter	Spring-flux	Pregnancy	Nursery	Mating	Autumn-flux
Total range	—	0–0	0–90	0–247	0–25	—
Reasonable likelihood	—	0–0	0–31	0–39.7	0–20.5	—
Significance	—	0	0	12.5	6	—

INTERNAL VOLUME	AVERAGE LENGTH OF BAT (top of head to rump) (mm)	AVERAGE WIDTH OF BAT (wrist to wrist) (mm)	AVERAGE CONDYLO-BASAL OF BAT (nose to back of skull) taken from Dietz *et al.* 2011 (mm)	AVERAGE VOLUME OF BAT (cm³)	TREE ROOST MATERNITY COLONY SIZE	MINIMUM VOLUME REQUIRED FOR A MATERNITY COLONY (cm³)
	69.5	35.5	24.1	59.5	20–130 (Harris and Yalden 2008)	1,190–7,735

APEX SHAPE	Winter	Spring-flux	Pregnancy	Nursery	Mating	Autumn-flux
Dome	—	—	44.4 (4)	45.5 (5)	62.5 (5)	—
Spire	—	100 (2)	55.6 (5)	36.4 (4)	37.5 (3)	—
Peak	—	—	11.1 (1)	27.3 (3)	—	—
Flat	—	—	—	—	—	—
Chambered	—	—	11.1 (1)	—	—	—
Tube	—	—	—	—	—	—

Table 8.9 Bechstein's bat *Myotis bechsteinii*: associated environment

HUMIDITY	Winter	Spring-flux	Pregnancy	Nursery	Mating	Autumn-flux
Dry	—	—	100 (8)	80 (8)	75 (3)	—
Damp	—	—	—	20 (2)	25 (1)	—
Wet	—	—	—	—	—	—

TEXTURE	Winter	Spring-flux	Pregnancy	Nursery	Mating	Autumn-flux
Smooth	—	—	87.5 (7)	77.8 (7)	100 (4)	—
Bobbly	—	—	25 (2)	22.2 (2)	25 (1)	—
Bumpy	—	—	37.5 (3)	22.2 (2)	—	—
Rough	—	—	25 (2)	22.2 (2)	—	—

CLEANLINESS	Winter	Spring-flux	Pregnancy	Nursery	Mating	Autumn-flux
Clean	—	—	87.5 (7)	88.9 (8)	75 (3)	—
Waxy	—	—	62.5 (5)	66.7 (6)	50 (2)	—
Blackened	—	—	50 (4)	55.6 (5)	50 (2)	—
Polished	—	—	25 (2)	33.3 (3)	—	—
Dirty	—	—	12.5 (1)	—	—	—
Dusty	—	—	12.5 (1)	—	—	—
Debris	—	—	12.5 (1)	—	—	—
Sludgy	—	—	12.5 (1)	11.1 (1)	25 (1)	—

ASSOCIATED COMPETITORS (i.e. species that may time-share PRF with Bechstein's bats)	Woodlice *Porcellio scaber*, tree slugs *Lehmannia marginata*, unidentified slugs, lipped snails *Cepaea* spp., spider's-webs, blue tits *Cyanistes caeruleus* and nuthatches *Sitta europaea*					

ODOUR	Winter	Spring-flux	Pregnancy	Nursery	Mating	Autumn-flux
None	—	50 (1)	66.7 (2)	20 (1)	66.7 (2)	—
Pleasant	—	—	33.3 (1)	—	—	—
Not unpleasant	—	50 (1)	—	80 (4)	33.3 (1)	—
Unpleasant	—	—	—	—	—	—

Table 8.10 Daubenton's bat *Myotis daubentonii*: associated dimensions (cm)

ENTRANCE HEIGHT	Winter	Spring-flux	Pregnancy	Nursery	Mating	Autumn-flux
Total range	—	5–52	5–325	1.2–143	3–35	52–52
Reasonable likelihood	—	8.5–20	9–40	5.8–61	5.5–22.5	52–52
Median	—	11.5	25	9	11.5	52
ENTRANCE WIDTH	Winter	Spring-flux	Pregnancy	Nursery	Mating	Autumn-flux
Total range	—	0.9–8	1.5–23	1.6–19	1–19	8–8
Reasonable likelihood	—	1.6–4	3.8–8	2.9–8	1.7–8.5	8–8
Median	—	4	5	5	3.5	8
INTERNAL HEIGHT	Winter	Spring-flux	Pregnancy	Nursery	Mating	Autumn-flux
Total range	—	12–121	12–431	20–121	22–101	121–121
Reasonable likelihood	—	26–71	38.5–72.5	25.8–69.2	43.7–94.2	121–121
Median	—	50	52.2	38	64.5	121
INTERNAL WIDTH	Winter	Spring-flux	Pregnancy	Nursery	Mating	Autumn-flux
Total range	—	2.5–14	2–12	1.2–12	2–12	6–6
Reasonable likelihood	—	5–8	4–8	5–11.5	4–10.25	6–6
Median	—	7	6.5	6.5	4.5	6
INTERNAL DEPTH	Winter	Spring-flux	Pregnancy	Nursery	Mating	Autumn-flux
Total range	—	0–318	0–450	0–318	0–100	318–318
Reasonable likelihood	—	0–4	0–0	0–5	0–0	318–318
Median	—	0	0	0	0	318
INTERNAL VOLUME	AVERAGE LENGTH OF BAT (top of head to rump) (mm)	AVERAGE WIDTH OF BAT (wrist to wrist) (mm)	AVERAGE CONDYLO-BASAL OF BAT (nose to back of skull) taken from Dietz *et al.* 2011 (mm)	AVERAGE VOLUME OF BAT (cm³)	TREE ROOST MATERNITY COLONY SIZE	MINIMUM VOLUME REQUIRED FOR A MATERNITY COLONY (cm³)
	72.5	37	20.1	53.8	20–50 (Dietz *et al.* 2011)	1,076–2,690
APEX SHAPE	Winter	Spring-flux	Pregnancy	Nursery	Mating	Autumn-flux
Dome	—	37.5 (3)	47.4 (9)	21.4 (3)	40 (2)	—
Spire	—	62.5 (5)	57.9 (11)	57.1 (8)	60 (3)	100 (1)
Peak	—	—	—	21.4 (3)	—	—
Flat	—	—	—	—	—	—
Chambered	—	—	10.5 (2)	28.6 (4)	20 (1)	—
Tube	—	—	—	—	—	—

Table 8.11 Daubenton's bat *Myotis daubentonii*: associated environment

HUMIDITY	Winter	Spring-flux	Pregnancy	Nursery	Mating	Autumn-flux
Dry	—	36.4 (4)	52.6 (10)	56.3 (9)	66.7 (4)	50 (1)
Damp	—	54.5 (6)	47.4 (9)	43.8 (7)	33.3 (2)	50 (1)
Wet	—	9.1 (1)	—	—	—	—

TEXTURE	Winter	Spring-flux	Pregnancy	Nursery	Mating	Autumn-flux
Smooth	—	42.9 (3)	50 (9)	46.2 (6)	66.7 (4)	100 (1)
Bobbly	—	57.1 (4)	38.9 (7)	61.5 (8)	33.3 (2)	100 (1)
Bumpy	—	—	11.1 (2)	53.8 (7)	33.3 (2)	—
Rough	—	28.6 (2)	27.8 (5)	15.4 (2)	16.7 (1)	—

CLEANLINESS	Winter	Spring-flux	Pregnancy	Nursery	Mating	Autumn-flux
Clean	—	42.9 (3)	39 (16)	47.8 (11)	33.3 (5)	100 (1)
Waxy	—	14.3 (1)	22 (9)	26.1 (6)	26.7 (4)	—
Blackened	—	28.6 (2)	22 (9)	17.4 (4)	20 (3)	—
Polished	—	14.3 (1)	17.1 (7)	8.7 (2)	20 (3)	—
Dirty	—	42.9 (3)	—	—	—	—
Dusty	—	14.3 (1)	—	—	—	—
Debris	—	14.3 (1)	—	—	—	—
Sludgy	—	28.6 (2)	5.9 (1)	8.3 (1)	—	—

ASSOCIATED COMPETITORS (i.e. species that may time-share PRF with Daubenton's bats)	Woodlice *Porcellio scaber*, tree slugs *Lehmannia marginata*, garden snails *Cornu aspersum*, lipped snails *Cepaea* spp., harvestmen *Leiobunum rotundum*, spider's-webs, spider egg-sacs, flies *Eustalomyia* (*anthomyiidae*), white-legged snake millipedes *Tachypodoiulus niger*, centipedes *Lithobius forficatus*, worms, blue tits *Cyanistes caeruleus*, unidentified birds (droppings and down)

ODOUR	Winter	Spring-flux	Pregnancy	Nursery	Mating	Autumn-flux
None	—	57.1 (4)	40 (4)	10 (1)	—	100 (1)
Pleasant	—	14.3 (1)	10 (1)	—	—	—
Not unpleasant	—	14.3 (1)	—	10 (1)	—	—
Unpleasant	—	14.3 (1)	50 (5)	80 (8)	100 (3)	—

Table 8.12 Natterer's bat *Myotis nattereri*: associated dimensions (cm)

ENTRANCE HEIGHT	Winter	Spring-flux	Pregnancy	Nursery	Mating	Autumn-flux
Total range	6–194	5–230	4–153	2–331	5–126	4.5–126
Reasonable likelihood	14.2–80	19.5–59.7	7–30.1	5.7–47	9.2–48	6–28.8
Median	29.5	27	15.5	20	29.4	13
ENTRANCE WIDTH	Winter	Spring-flux	Pregnancy	Nursery	Mating	Autumn-flux
Total range	2.7–6.2	1–7	1–20	1.5–15	1.5–20	1–15
Reasonable likelihood	3.3–5.3	1.8–4.5	1.9–5.1	2.9–6.2	2–5.3	2–6.5
Median	4.25	3.7	3	4	3.25	3
INTERNAL HEIGHT	Winter	Spring-flux	Pregnancy	Nursery	Mating	Autumn-flux
Total range	4–119	2.5–90	2.5–126	4–98	0–122	2.5–121
Reasonable likelihood	25.3–54.5	19–35	22.1–60	25.1–42.7	20.2–45.7	18.3–44.5
Median	32.75	24	34	32.3	32.75	21.5
INTERNAL WIDTH	Winter	Spring-flux	Pregnancy	Nursery	Mating	Autumn-flux
Total range	2–40	1.8–28	2.5–20	2–15	1.5–150	2–18
Reasonable likelihood	3.1–31	3–6	3.2–10.4	3.1–8	4–6	2.5–6
Median	15.75	4.6	5	5	5	3
INTERNAL DEPTH	Winter	Spring-flux	Pregnancy	Nursery	Mating	Autumn-flux
Total range	0–0	0–67	0–101	0–331	0–318	0–318
Reasonable likelihood	0–0	0–5	0–10	0–6.1	0–3.5	0–0
Median	0	0	0	0	0	0

INTERNAL VOLUME	AVERAGE LENGTH OF BAT (top of head to rump) (mm)	AVERAGE WIDTH OF BAT (wrist to wrist) (mm)	AVERAGE CONDYLO-BASAL OF BAT (nose to back of skull) taken from Dietz *et al.* 2011 (mm)	AVERAGE VOLUME OF BAT (cm³)	TREE ROOST MATERNITY COLONY SIZE	MINIMUM VOLUME REQUIRED FOR A MATERNITY COLONY (cm³)
	65.0	32.5	21.6	45.6	20–50 (Harris and Yalden 2008; Dietz *et al.* 2011)	912–2,280

APEX SHAPE	Winter	Spring-flux	Pregnancy	Nursery	Mating	Autumn-flux
Dome	50 (2)	12.5 (3)	27.3 (6)	13.6 (3)	4.5 (1)	25 (4)
Spire	50 (2)	66.7 (16)	54.5 (12)	54.5 (12)	50 (11)	68.8 (11)
Peak	—	20.8 (5)	22.7 (5)	31.8 (7)	27.3 (6)	6.3 (1)
Flat	—	—	—	—	13.6 (3)	—
Chambered	—	8.3 (2)	22.7 (5)	27.3 (6)	9.1 (2)	12.5 (2)
Tube	—	—	—	—	9.1 (2)	—

Table 8.13 Table 55. Natterer's bat *Myotis nattereri*: associated environment

HUMIDITY	Winter	Spring-flux	Pregnancy	Nursery	Mating	Autumn-flux
Dry	66.7 (2)	68.0 (17)	87.5 (21)	90.9 (20)	61.1 (11)	61.1 (11)
Damp	33.3 (1)	24 (6)	12.5 (3)	4.5 (1)	38.9 (7)	33.3 (6)
Wet	—	8 (2)	—	4.5 (1)	—	5.6 (1)

TEXTURE	Winter	Spring-flux	Pregnancy	Nursery	Mating	Autumn-flux
Smooth	—	65.2 (15)	78.3 (18)	81.8 (18)	56.3 (9)	69.2 (9)
Bobbly	33.3 (1)	34.8 (8)	13.0 (3)	18.2 (4)	18.8 (3)	46.2 (6)
Bumpy	66.7 (2)	21.7 (5)	39.1 (9)	40.9 (9)	25.0 (4)	23.1 (3)
Rough	66.7 (2)	21.7 (5)	39.1 (9)	40.9 (9)	25.0 (4)	15.4 (2)

CLEANLINESS	Winter	Spring-flux	Pregnancy	Nursery	Mating	Autumn-flux
Clean	100 (3)	95.2 (20)	95.5 (21)	90.9 (20)	100 (16)	92.3 (12)
Waxy	66.7 (2)	52.4 (11)	63.6 (14)	77.3 (17)	43.8 (7)	53.8 (7)
Blackened	—	33.3 (7)	40.9 (9)	77.3 (17)	37.5 (6)	30.8 (4)
Polished	—	23.8 (5)	22.7 (5)	63.6 (14)	31.3 (5)	30.8 (4)
Dirty	—	5 (1)	—	4.5 (1)	—	7.7 (1)
Dusty	—	—	4.5 (1)	—	—	—
Debris	—	—	4.5 (1)	—	—	—
Sludgy	—	5 (1)	—	4.5 (1)	6.3 (1)	7.7 (1)

ASSOCIATED COMPETITORS (i.e. species that may time-share PRF with Natterer's bats)	Woodlice *Porcellio scaber*, dusky slugs *Arion subfuscus*, leopard slugs *Limax maximus*, unidentified slugs, garden snails *Cornu aspersum*, lipped snails *Cepaea* spp., unidentified snails, harvestmen *Leiobunum rotundum*, unidentified spiders, spider's-webs, spider egg-sacs, unidentified flies, tree-wasp *Vespula sylvestris* nests, hornet *Vespa crabro* (individual), unidentified beetles, millipedes *Ophyiulus pilosus*, centipedes *Lithobius forficatus*, worms, copper underwing *Amphipyra pyramidea*, blue tits *Cyanistes caeruleus*, unidentified birds (droppings and down) and common dormice *Muscardinus avellanarius* (nests)

ODOUR	Winter	Spring-flux	Pregnancy	Nursery	Mating	Autumn-flux
None	100 (3)	80 (12)	42.9 (9)	50 (10)	57.1 (12)	76.9 (10)
Pleasant	—	—	14.3 (3)	5 (1)	14.3 (3)	15.4 (2)
Not unpleasant	—	20 (3)	33.3 (7)	45 (9)	23.8 (5)	15.4 (2)
Unpleasant	—	—	9.5 (2)	—	4.8 (1)	—

Table 8.14 Noctule *Nyctalus noctula*: associated dimensions (cm)

ENTRANCE HEIGHT	Winter	Spring-flux	Pregnancy	Nursery	Mating	Autumn-flux
Total range	0–169	5.5–227	4.5–145	1.2–15	7–227	6.5–201
Reasonable likelihood	8.8–94	8.7–111.2	7–10.6	5.8–11.2	11.05–169.2	9.5–117.5
Median	29.5	22.5	8	7	84.5	22.5
ENTRANCE WIDTH	Winter	Spring-flux	Pregnancy	Nursery	Mating	Autumn-flux
Total range	1.8–20	1.8–20	4.6–8.9	6–11	2.5–5.5	1.6–20
Reasonable likelihood	2.5–6.2	5–7.3	5.5–6.5	6.3–9.1	4.6–5.5	2–6.3
Median	5.8	5.8	6.3	7.2	5	3
INTERNAL HEIGHT	Winter	Spring-flux	Pregnancy	Nursery	Mating	Autumn-flux
Total range	21–180	6–180	0–88.1	0–90	20–115	22–180
Reasonable likelihood	24.3–59	21.5–67	19.4–40	0–85.7	31.2–56	39.2–100.7
Median	36.5	31.7	21	44	37	63.7
INTERNAL WIDTH	Winter	Spring-flux	Pregnancy	Nursery	Mating	Autumn-flux
Total range	2.1–150	4–150	8–50	1.2–50	4–20	2–150
Reasonable likelihood	4.7–9.2	7.3–18.7	11–21	19.1–41.2	5.3–12.3	4.6–7.7
Median	6	9.5	15	23.1	7.5	5
INTERNAL DEPTH	Winter	Spring-flux	Pregnancy	Nursery	Mating	Autumn-flux
Total range	0–267	0–267	0–51	0–50	0–28.4	0–267
Reasonable likelihood	0–10	0–12.7	2–19.1	15.5–42	0–12.2	0–11.7
Median	0	3	6.5	23	9	1.2
INTERNAL VOLUME	AVERAGE LENGTH OF BAT (top of head to rump) (mm)	AVERAGE WIDTH OF BAT (wrist to wrist) (mm)	AVERAGE CONDYLO-BASAL OF BAT (nose to back of skull) taken from Dietz *et al.* 2011 (mm)	AVERAGE VOLUME OF BAT (cm³)	TREE ROOST MATERNITY COLONY SIZE	MINIMUM VOLUME REQUIRED FOR A MATERNITY COLONY (cm³)
	101	47	11.2	52.9	10–95 (BTHK Database)	1,059–5,025
APEX SHAPE	Winter	Spring-flux	Pregnancy	Nursery	Mating	Autumn-flux
Dome	33.3 (4)	41.7 (5)	100 (9)	50 (4)	12.5 (1)	40 (4)
Spire	25 (3)	16.7 (2)	—	—	25 (2)	50 (5)
Peak	41.7 (5)	16.7 (2)	—	37.5 (3)	25 (2)	20 (2)
Flat	—	8.3 (1)	—	12.5 (1)	12.5 (1)	—
Chambered	16.7 (2)	8.3 (1)	—	12.5 (1)	12.5 (1)	20 (2)
Tube	—	8.3 (1)	—	—	25 (2)	—

Table 8.15 Noctule *Nyctalus noctula*: associated environment

HUMIDITY	Winter	Spring-flux	Pregnancy	Nursery	Mating	Autumn-flux
Dry	90 (9)	100 (6)	50 (4)	80 (8)	100 (7)	81.8 (9)
Damp	—	—	50 (4)	20 (2)	—	18.2 (2)
Wet	10 (1)	—	—	—	—	—

TEXTURE	Winter	Spring-flux	Pregnancy	Nursery	Mating	Autumn-flux
Smooth	88.9 (8)	100 (6)	57.1 (4)	100 (8)	87.5 (7)	55.6 (5)
Bobbly	11.1 (1)	16.7 (1)	42.9 (3)	12.5 (1)	—	22.2 (2)
Bumpy	22.2 (2)	50 (3)	71.4 (5)	50 (4)	25 (2)	44.4 (4)
Rough	22.2 (2)	—	14.3 (1)	—	—	—

CLEANLINESS	Winter	Spring-flux	Pregnancy	Nursery	Mating	Autumn-flux
Clean	100 (9)	100 (6)	85.7 (6)	100 (8)	100 (8)	88.9 (8)
Waxy	66.7 (6)	83.3 (5)	57.1 (4)	100 (8)	100 (8)	77.8 (7)
Blackened	77.8 (7)	66.7 (4)	85.7 (6)	75 (6)	87.5 (7)	33.3 (3)
Polished	11.1 (1)	16.7 (1)	14.3 (1)	12.5 (1)	37.5 (3)	—
Dirty	—	—	14.3 (1)	—	—	—
Dusty	11.1 (1)	—	—	—	—	—
Debris	—	—	—	—	—	11.1 (1)
Sludgy	—	—	28.6 (2)	12.5 (1)	—	11.1 (1)

ASSOCIATED COMPETITORS (i.e. species that may time-share PRF with noctules)	Woodlice *Porcellio scaber*, spider's-webs, unidentified flies, wood ants *Formica* sp., nuthatches *Sitta europaea* and unidentified birds (nests, droppings and down)					

ODOUR	Winter	Spring-flux	Pregnancy	Nursery	Mating	Autumn-flux
None	33.3 (2)	11.1 (1)	—	—	14.3 (1)	12.5 (1)
Pleasant	—	—	—	—	—	12.5 (1)
Not unpleasant	33.3 (2)	55.6 (5)	33.3 (3)	—	57.1 (4)	25 (2)
Unpleasant	33.3 (2)	33.3 (3)	66.7 (6)	100 (7)	28.6 (2)	50 (4)

Table 8.16 Nathusius' pipistrelle *Pipistrellus nathusii*: associated dimensions (cm)

ENTRANCE HEIGHT	Winter	Spring-flux	Pregnancy	Nursery	Mating	Autumn-flux
Total range	—	—	—	—	—	23.5–23.5
Reasonable likelihood	—	—	—	—	—	23.5–23.5
Median	—	—	—	—	—	23.5
ENTRANCE WIDTH	Winter	Spring-flux	Pregnancy	Nursery	Mating	Autumn-flux
Total range	—	—	—	—	—	2–2
Reasonable likelihood	—	—	—	—	—	2–2
Median	—	—	—	—	—	2
INTERNAL HEIGHT	Winter	Spring-flux	Pregnancy	Nursery	Mating	Autumn-flux
Total range	—	—	—	—	—	13–13
Reasonable likelihood	—	—	—	—	—	13–13
Median	—	—	—	—	—	13
INTERNAL WIDTH	Winter	Spring-flux	Pregnancy	Nursery	Mating	Autumn-flux
Total range	—	—	—	—	—	6–6
Reasonable likelihood	—	—	—	—	—	6–6
Median	—	—	—	—	—	6
INTERNAL DEPTH	Winter	Spring-flux	Pregnancy	Nursery	Mating	Autumn-flux
Total range	—	—	—	—	—	0–0
Reasonable likelihood	—	—	—	—	—	0–0
Median	—	—	—	—	—	0

INTERNAL VOLUME	AVERAGE LENGTH OF BAT (top of head to rump) (mm)	AVERAGE WIDTH OF BAT (wrist to wrist) (mm)	AVERAGE CONDYLO-BASAL OF BAT (nose to back of skull) taken from Dietz *et al.* 2011 (mm)	AVERAGE VOLUME OF BAT (cm³)	TREE ROOST MATERNITY COLONY SIZE	MINIMUM VOLUME REQUIRED FOR A MATERNITY COLONY (cm³)
	73.5	37	19.2	52.1	No data	—

APEX SHAPE	Winter	Spring-flux	Pregnancy	Nursery	Mating	Autumn-flux
Dome	—	—	—	—	—	—
Spire	—	—	—	—	—	100 (1)
Peak	—	—	—	—	—	—
Flat	—	—	—	—	—	—
Chambered	—	—	—	—	—	100 (1)
Tube	—	—	—	—	—	—

Table 8.17 Nathusius' pipistrelle *Pipistrellus nathusii*: associated environment

HUMIDITY	Winter	Spring-flux	Pregnancy	Nursery	Mating	Autumn-flux
Dry	—	—	—	—	—	100 (1)
Damp	—	—	—	—	—	—
Wet	—	—	—	—	—	—

TEXTURE	Winter	Spring-flux	Pregnancy	Nursery	Mating	Autumn-flux
Smooth	—	—	—	—	—	—
Bobbly	—	—	—	—	—	—
Bumpy	—	—	—	—	—	—
Rough	—	—	—	—	—	100 (1)

CLEANLINESS	Winter	Spring-flux	Pregnancy	Nursery	Mating	Autumn-flux
Clean	—	—	—	—	—	—
Waxy	—	—	—	—	—	100 (1)
Blackened	—	—	—	—	—	100 (1)
Polished	—	—	—	—	—	—
Dirty	—	—	—	—	—	—
Dusty	—	—	—	—	—	100 (1)
Debris	—	—	—	—	—	100 (1)
Sludgy	—	—	—	—	—	—

ASSOCIATED COMPETITORS (i.e. species that may time-share PRF with Nathusius' pipistrelles)	Spider's-webs					

ODOUR	Winter	Spring-flux	Pregnancy	Nursery	Mating	Autumn-flux
None	—	—	—	—	—	100 (1)
Pleasant	—	—	—	—	—	—
Not unpleasant	—	—	—	—	—	—
Unpleasant	—	—	—	—	—	—

Table 8.18 Common pipistrelle *Pipistrellus pipistrellus*: associated dimensions (cm)

ENTRANCE HEIGHT	Winter	Spring-flux	Pregnancy	Nursery	Mating	Autumn-flux
Total range	1.5–160	1.5–152	4.2–132	6.2–250	0–121	1.5–510
Reasonable likelihood	11–74	6.2–79.5	46.8–85	7.5–39.8	3.5–31.5	8.7–124.5
Median	30	36.5	70	19.7	7	28.2

ENTRANCE WIDTH	Winter	Spring-flux	Pregnancy	Nursery	Mating	Autumn-flux
Total range	1–4.9	1–10	2.5–10	1.5–15	2–11	1–19.7
Reasonable likelihood	1.5–2	1.5–2.9	5–7.5	2–7	2–6.7	2–4.9
Median	1.8	1.9	6.2	5.4	2.6	2.2

INTERNAL HEIGHT	Winter	Spring-flux	Pregnancy	Nursery	Mating	Autumn-flux
Total range	0–65	0–55	0–35	6–60	11–50	0–90
Reasonable likelihood	0–32.5	16.2–35.1	12.7–28.9	15.75–3	18–40	13–34
Median	20	26	20	22.8	27.5	25

INTERNAL WIDTH	Winter	Spring-flux	Pregnancy	Nursery	Mating	Autumn-flux
Total range	2.5–49.5	2–20	3.2–40	2–10.1	2.5–40	1.5–49.5
Reasonable likelihood	4–18	3.8–17.6	8.5–13.2	2.8–5	4.2–11.7	2.5–6
Median	10	10	10	3.2	8	3

INTERNAL DEPTH	Winter	Spring-flux	Pregnancy	Nursery	Mating	Autumn-flux
Total range	0–26	0–18	0–42.5	0–26	0–182	0–5
Reasonable likelihood	0–0	0–0	0–10	0–0	0–12.2	0–0
Median	0	0	0	0	2	0

INTERNAL VOLUME	AVERAGE LENGTH OF BAT (top of head to rump) (mm)	AVERAGE WIDTH OF BAT (wrist to wrist) (mm)	AVERAGE CONDYLO-BASAL OF BAT (nose to back of skull) taken from Dietz *et al.* 2011 (mm)	AVERAGE VOLUME OF BAT (cm³)	TREE ROOST MATERNITY COLONY SIZE	MINIMUM VOLUME REQUIRED FOR A MATERNITY COLONY (cm³)
	57	30	16.4	28	36 (Howe 1997)	1,007

APEX SHAPE	Winter	Spring-flux	Pregnancy	Nursery	Mating	Autumn-flux
Dome	16.7 (2)	16.7 (2)	—	12.5 (1)	50 (3)	23.1 (3)
Spire	16.7 (2)	16.7 (2)	—	25 (2)	16.7 (1)	15.4 (2)
Peak	66.7 (8)	66.7 (8)	66.7 (4)	50 (4)	33.3 (2)	53.8 (7)
Flat	—	—	16.7 (1)	12.5 (1)	—	7.7 (1)
Chambered	—	—	—	—	33.3 (2)	30.8 (4)
Tube	—	—	—	—	—	7.7 (1)

Table 8.19 Common pipistrelle *Pipistrellus pipistrellus*: associated environment

HUMIDITY	Winter	Spring-flux	Pregnancy	Nursery	Mating	Autumn-flux
Dry	100 (10)	100 (11)	77.8 (7)	100 (6)	100 (6)	100 (12)
Damp	—	—	22.2 (2)	—	—	—
Wet	—	—	—	—	—	—
TEXTURE	**Winter**	**Spring-flux**	**Pregnancy**	**Nursery**	**Mating**	**Autumn-flux**
Smooth	80 (8)	72.7 (8)	28.6 (2)	50 (3)	50 (3)	38.5 (5)
Bobbly	20 (2)	—	—	16.7 (1)	—	7.7 (1)
Bumpy	10 (1)	36.4 (4)	—	33.3 (2)	50 (3)	53.8 (7)
Rough	30 (3)	45.5 (5)	71.4 (5)	50 (3)	66.7 (4)	61.5 (8)
CLEANLINESS	**Winter**	**Spring-flux**	**Pregnancy**	**Nursery**	**Mating**	**Autumn-flux**
Clean	90 (9)	81.8 (9)	100 (7)	100 (6)	100 (6)	83.3 (10)
Waxy	30 (3)	45.5 (5)	14.3 (1)	33.3 (2)	66.7 (4)	41.7 (5)
Blackened	20 (2)	36.4 (4)	—	—	16.7 (1)	16.7 (2)
Polished	—	9.1 (1)	—	—	—	8.3 (1)
Dirty	—	—	—	—	—	—
Dusty	—	27.3 (3)	14.3 (1)	—	—	8.3 (1)
Debris	20 (2)	27.3 (3)	28.6 (2)	—	—	16.7 (2)
Sludgy	—	—	—	—	—	—
ASSOCIATED COMPETITORS (i.e. species that may time-share PRF with common pipistrelles)	Woodlice *Porcellio scaber*, tree slugs *Lehmannia marginata*, garden snails *Cornu aspersum*, unidentified spiders and spider's-webs					
ODOUR	**Winter**	**Spring-flux**	**Pregnancy**	**Nursery**	**Mating**	**Autumn-flux**
None	80 (8)	90 (9)	75 (3)	50 (3)	80 (4)	63.6 (7)
Pleasant	10 (1)	—	25 (1)	—	—	9.1 (1)
Not unpleasant	10 (1)	10 (1)	—	50 (3)	—	9.1 (1)
Unpleasant	—	—	—	—	20 (1)	18.2 (2)

Table 8.20 Soprano pipistrelle *Pipistrellus pygmaeus*: associated dimensions (cm)

ENTRANCE HEIGHT	Winter	Spring-flux	Pregnancy	Nursery	Mating	Autumn-flux
Total range	1.5–23.5	16–121	2.8–200	17–100	6–12	1.5–102
Reasonable likelihood	5.5–16.5	21.6–47.8	3.1–22.2	25–66.5	6.1–9.1	4.8–60
Median	9.5	23.5	11.7	33	6.2	26
ENTRANCE WIDTH	Winter	Spring-flux	Pregnancy	Nursery	Mating	Autumn-flux
Total range	1.5–6	1.5–2	2–9	1.3–5	2–5.4	1.2–3
Reasonable likelihood	1.7–4	1.8–2	2.7–7.5	1.6–3.5	2.2–3.9	1.4–2.4
Median	2	2	5.2	2	2.5	1.8
INTERNAL HEIGHT	Winter	Spring-flux	Pregnancy	Nursery	Mating	Autumn-flux
Total range	0–13	11–101	10–69	8–34	2–20	0–58
Reasonable likelihood	6–12.5	12.5–35	14.7–58	19–32	9.5–18.5	0–22
Median	12	13	32.7	30	17	5
INTERNAL WIDTH	Winter	Spring-flux	Pregnancy	Nursery	Mating	Autumn-flux
Total range	6–20	4–6	1.7–26	1.3–40	2.5–3.2	1.7–20
Reasonable likelihood	8–15	4.7–6	2.2–5.7	1.9–21.2	2.7–3.1	2.3–11.7
Median	10	5.5	2.9	2.5	3	5.7
INTERNAL DEPTH	Winter	Spring-flux	Pregnancy	Nursery	Mating	Autumn-flux
Total range	0–0	0–0	0–10	0–0	0–0	0–0
Reasonable likelihood	0–0	0–0	0–0	0–0	0–0	0–0
Median	0	0	0	0	0	0

INTERNAL VOLUME	AVERAGE LENGTH OF BAT (top of head to rump) (mm)	AVERAGE WIDTH OF BAT (wrist to wrist) (mm)	AVERAGE CONDYLO-BASAL OF BAT (nose to back of skull) taken from Dietz *et al.* 2011 (mm)	AVERAGE VOLUME OF BAT (cm³)	TREE ROOST MATERNITY COLONY SIZE	MINIMUM VOLUME REQUIRED FOR A MATERNITY COLONY (cm³)
	57	28	16.4	26.2	No data	—

APEX SHAPE	Winter	Spring-flux	Pregnancy	Nursery	Mating	Autumn-flux
Dome	33.3 (1)	25 (1)	16.7 (1)	33.3 (1)	33.3 (1)	25 (1)
Spire	66.7 (2)	75 (3)	33.3 (2)	—	—	50 (2)
Peak	—	—	50 (3)	66.7 (2)	33.3 (1)	25 (1)
Flat	—	—	—	—	33.3 (1)	—
Chambered	33.3 (1)	50 (2)	—	—	—	25 (1)
Tube	—	25 (1)	—	—	—	—

Table 8.21 Soprano pipistrelle *Pipistrellus pygmaeus*: associated environment

HUMIDITY	Winter	Spring-flux	Pregnancy	Nursery	Mating	Autumn-flux
Dry	100 (2)	100 (4)	83.3 (5)	100 (3)	100 (1)	75 (3)
Damp	—	—	16.7 (1)	—	—	25 (1)
Wet	—	—	—	—	—	—
TEXTURE	**Winter**	**Spring-flux**	**Pregnancy**	**Nursery**	**Mating**	**Autumn-flux**
Smooth	—	50 (2)	80 (4)	66.7 (2)	100 (1)	33.3 (1)
Bobbly	—	—	20 (1)	—	—	—
Bumpy	—	25 (1)	60 (3)	—	—	—
Rough	**100 (2)**	**100 (4)**	—	33.3 (1)	—	66.7 (2)
CLEANLINESS	**Winter**	**Spring-flux**	**Pregnancy**	**Nursery**	**Mating**	**Autumn-flux**
Clean	—	50 (2)	100 (6)	100 (3)	100 (1)	100 (3)
Waxy	50 (1)	100 (4)	33.3 (2)	33.3 (1)	—	33.3 (1)
Blackened	50 (1)	100 (4)	33.3 (2)	—	—	—
Polished	—	—	—	—	—	—
Dirty	—	—	—	—	—	33.3 (1)
Dusty	50 (1)	50 (2)	—	—	—	33.3 (1)
Debris	50 (1)	75 (3)	—	—	—	33.3 (1)
Sludgy	—	—	—	—	—	—
ASSOCIATED COMPETITORS (i.e. species that may time-share PRF with soprano pipistrelles)	Woodlice *Porcellio scaber*, dusky slugs *Arion subfuscus*, unidentified spiders, spider's-webs and unidentified bees (individual)					
ODOUR	**Winter**	**Spring-flux**	**Pregnancy**	**Nursery**	**Mating**	**Autumn-flux**
None	100 (2)	100 (4)	—	50 (1)	100 (2)	—
Pleasant	—	—	—	—	—	—
Not unpleasant	—	—	—	50 (1)	—	100 (1)
Unpleasant	—	—	100 (3)	—	—	—

Table 8.22 Brown long-eared bat *Plecotus auritus*: associated dimensions (cm)

ENTRANCE HEIGHT	Winter	Spring-flux	Pregnancy	Nursery	Mating	Autumn-flux
Total range	1.5–208	2–223	5.5–159	3.4–150	0–227	3.4–387
Reasonable likelihood	12.5–99	12–89	8–20	7.7–48.3	10–69	10.6–89
Median	32	29	13	15.5	27	26.5
ENTRANCE WIDTH	Winter	Spring-flux	Pregnancy	Nursery	Mating	Autumn-flux
Total range	1–20	1–25	1–20	2–8.6	0–10	0.8–30
Reasonable likelihood	2–6	2.5–6	3–3.5	2.6–5.5	2.5–5	2.1–6
Median	4	4	3	4	3	3.2
INTERNAL HEIGHT	Winter	Spring-flux	Pregnancy	Nursery	Mating	Autumn-flux
Total range	0–95	0–123	2–121	0–121	0–190	0–126
Reasonable likelihood	20–63.5	15.2–52.7	21–47.8	23.5–55.2	24.7–60.7	19.1–52.8
Median	43	29	35	39.5	35	30.5
INTERNAL WIDTH	Winter	Spring-flux	Pregnancy	Nursery	Mating	Autumn-flux
Total range	1.5–35	0–27	2–20	3–23.6	2–28	0–18
Reasonable likelihood	3–6	3.5–6	4–10.3	4.2–12.7	4–8	3.5–6
Median	4.5	5	5.2	7.5	5	5
INTERNAL DEPTH	Winter	Spring-flux	Pregnancy	Nursery	Mating	Autumn-flux
Total range	0–214	0–99	0–318	0–318	0–34	0–318
Reasonable likelihood	0–3	0–3.3	0–10	0–17.4	0–0	0–0
Median	0	0	0	2.5	0	0

INTERNAL VOLUME	AVERAGE LENGTH OF BAT (top of head to rump) (mm)	AVERAGE WIDTH OF BAT (wrist to wrist) (mm)	AVERAGE CONDYLO-BASAL OF BAT (nose to back of skull) taken from Dietz *et al.* 2011 (mm)	AVERAGE VOLUME OF BAT (cm3)	TREE ROOST MATERNITY COLONY SIZE	MINIMUM VOLUME REQUIRED FOR A MATERNITY COLONY (cm3)
	63	32	22.1	44.6	7–28 (BTHK Database)	312–1,249

APEX SHAPE	Winter	Spring-flux	Pregnancy	Nursery	Mating	Autumn-flux
Dome	30.3 (10)	23 (23)	27.3 (6)	35.7 (5)	25 (8)	16.7 (9)
Spire	39.4 (13)	42 (42)	18.2 (4)	28.6 (4)	25 (8)	53.7 (29)
Peak	24.2 (8)	36 (36)	45.5 (10)	21.4 (3)	28.1 (9)	24.1 (13)
Flat	6.1 (2)	3 (3)	9.1 (2)	—	—	3.7 (2)
Chambered	9.1 (3)	16 (16)	13.6 (3)	14.3 (2)	6.3 (2)	18.5 (10)
Tube	3 (1)	2 (2)	4.5 (1)	—	15.6 (5)	—

Table 8.23 Brown long-eared bat *Plecotus auritus*: associated environment

HUMIDITY	Winter	Spring-flux	Pregnancy	Nursery	Mating	Autumn-flux
Dry	82.4 (28)	88.9 (80)	91.3 (21)	75 (9)	84.8 (28)	78.2 (43)
Damp	14.7 (5)	10 (9)	8.7 (2)	16.7 (2)	12.1 (4)	20 (11)
Wet	2.9 (1)	1.1 (1)	-	8.3 (1)	3 (1)	1.8 (1)
TEXTURE	Winter	Spring-flux	Pregnancy	Nursery	Mating	Autumn-flux
Smooth	68.8 (22)	68.3 (56)	91.3 (21)	58.3 (7)	74.2 (23)	47.1 (24)
Bobbly	21.9 (7)	13.4 (11)	4.3 (1)	16.7 (2)	16.1 (5)	19.6 (10)
Bumpy	46.9 (15)	24.4 (20)	30.4 (7)	33.3 (4)	22.6 (7)	45.1 (23)
Rough	37.5 (12)	36.6 (30)	13 (3)	25 (3)	22.6 (7)	54.9 (28)
CLEANLINESS	Winter	Spring-flux	Pregnancy	Nursery	Mating	Autumn-flux
Clean	93.8 (30)	95.1 (78)	91.3 (21)	75 (9)	90.3 (28)	86.3 (44)
Waxy	46.9 (15)	51.2 (42)	60.9 (14)	58.3 (7)	51.6 (16)	43.1 (22)
Blackened	25 (8)	31.7 (26)	34.8 (8)	50 (6)	45.2 (14)	27.5 (14)
Polished	25 (8)	14.6 (12)	30.4 (7)	33.3 (4)	19.4 (6)	5.9 (3)
Dirty	3.1 (1)	1.2 (1)	8.7 (2)	16.7 (2)	6.5 (2)	7.8 (4)
Dusty	3.1 (1)	3.7 (3)	8.7 (2)	8.3 (1)	6.5 (2)	5.9 (3)
Debris	6.3 (2)	4.9 (4)	13 (3)	250 (3)	6.5 (2)	19.6 (10)
Sludgy	—	2.5 (2)	4.3 (1)	8.3 (1)	—	2 (1)
ASSOCIATED COMPETITORS (i.e. species that may time-share PRF with brown long-eared bats)	Woodlice *Porcellio scaber*, dusky slugs *Arion subfuscus*, tree slugs *Lehmannia marginata*, unidentified slugs, garden snails *Cornu aspersum*, lipped snails *Cepaea* spp., unidentified snails, harvestmen *Leiobunum rotundum*, unidentified spiders, spider's-webs, spider egg-sacs, unidentified flies, millipedes *Ophyiulus pilosus*, centipedes *Lithobius forficatus*, worms, blue tits *Cyanistes caeruleus*, unidentified birds (nests, droppings and down), grey squirrels *Sciurus carolinensis* and common dormice *Muscardinus avellanarius* (nests)					
ODOUR	Winter	Spring-flux	Pregnancy	Nursery	Mating	Autumn-flux
None	88.5 (23)	72.9 (62)	47.4 (9)	25 (2)	52.2 (12)	72.3 (34)
Pleasant	3.8 (1)	5.9 (5)	15.8 (3)	—	—	14.9 (7)
Not unpleasant	3.8 (1)	21.2 (18)	21.1 (4)	50 (4)	30.4 (7)	12.8 (6)
Unpleasant	3.8 (1)	—	15.8 (3)	25 (2)	17.4 (4)	—

Table 8.24 Lesser horseshoe-bat *Rhinolophus hipposideros*: associated dimensions (cm)

ENTRANCE HEIGHT	Winter	Spring-flux	Pregnancy	Nursery	Mating	Autumn-flux
Total range	—	—	68–68	—	86–86	86–194
Reasonable likelihood	—	—	68–68	—	86–86	113–167
Median	—	—	68	—	86	140
ENTRANCE WIDTH	Winter	Spring-flux	Pregnancy	Nursery	Mating	Autumn-flux
Total range	—	—	35–35	—	37–37	35–37
Reasonable likelihood	—	—	35–35	—	37–37	35.5–36.5
Median	—	—	35	—	37	36
INTERNAL HEIGHT	Winter	Spring-flux	Pregnancy	Nursery	Mating	Autumn-flux
Total range	—	—	201–201	—	227–227	227–441
Reasonable likelihood	—	—	201–201	—	227–227	280.5–387.5
Median	—	—	201	—	227	334
INTERNAL WIDTH	Winter	Spring-flux	Pregnancy	Nursery	Mating	Autumn-flux
Total range	—	—	85–85	—	63–63	53–63
Reasonable likelihood	—	—	85–85	—	63–63	55.5–60.5
Median	—	—	85	—	63	58
INTERNAL DEPTH	Winter	Spring-flux	Pregnancy	Nursery	Mating	Autumn-flux
Total range	—	—	57–57	—	146–146	0–146
Reasonable likelihood	—	—	57–57	—	146–146	36.5–109.5
Median	—	—	57	—	146	73
INTERNAL VOLUME	AVERAGE LENGTH OF BAT (top of head to rump) (mm)	AVERAGE WIDTH OF BAT (wrist to wrist) (mm)	AVERAGE CONDYLO-BASAL OF BAT (nose to back of skull) taken from Dietz *et al.* 2011 (mm)	AVERAGE VOLUME OF BAT (cm³)	TREE ROOST MATERNITY COLONY SIZE	MINIMUM VOLUME REQUIRED FOR A MATERNITY COLONY (cm³)
	N/A	N/A	N/A	N/A	N/A	N/A
APEX SHAPE	Winter	Spring-flux	Pregnancy	Nursery	Mating	Autumn-flux
Dome	—	—	100 (1)	—	100 (1)	100 (2)
Spire	—	—	—	—	—	—
Peak	—	—	—	—	—	—
Flat	—	—	—	—	—	—
Chambered	—	—	100 (1)	—	—	—
Tube	—	—	—	—	—	—

Table 8.25 Lesser horseshoe-bat *Rhinolophus hipposideros*: associated environment

HUMIDITY	Winter	Spring-flux	Pregnancy	Nursery	Mating	Autumn-flux
Dry	—	—	100 (1)	—	—	33.3 (1)
Damp	—	—	—	—	100 (1)	66.7 (2)
Wet	—	—	—	—	—	—

TEXTURE	Winter	Spring-flux	Pregnancy	Nursery	Mating	Autumn-flux
Smooth	—	—	—	—	—	—
Bobbly	—	—	—	—	—	—
Bumpy	—	—	—	—	—	—
Rough	—	—	100 (1)	—	100 (1)	100 (2)

CLEANLINESS	Winter	Spring-flux	Pregnancy	Nursery	Mating	Autumn-flux
Clean	—	—	—	—	100 (1)	50 (1)
Waxy	—	—	—	—	—	—
Blackened	—	—	—	—	—	—
Polished	—	—	—	—	—	—
Dirty	—	—	—	—	—	—
Dusty	—	—	100 (1)	—	100 (1)	100 (2)
Debris	—	—	—	—	100 (1)	100 (2)
Sludgy	—	—	—	—	—	—

ASSOCIATED COMPETITORS (i.e. species that may time-share PRF with lesser horseshoe-bats)	Cave spider's *Meta menardi* and spider's webs

ODOUR	Winter	Spring-flux	Pregnancy	Nursery	Mating	Autumn-flux
None	—	—	—	—	—	100 (2)
Pleasant	—	—	—	—	—	—
Not unpleasant	—	—	—	—	100 (1)	—
Unpleasant	—	—	—	—	—	—

8.5.7 Summing-up

The final analysis is as follows:

» If bats, droppings and/or bat-flies are encountered, proceed to Scenario 1.

» If no conclusive evidence is recorded but the dimensions and environmental conditions are *all* within the range of *"reasonable likelihood"* for a specific bat species within a particular period of the year, and the internal volume is sufficient, proceed to Scenario 2.

» If the dimensions and environmental conditions fall within either a mix of *"reasonable likelihood"* and the total range, or solely within the total range, for a specific bat species within a particular period, and the internal volume is sufficient, proceed to Scenario 3.

» If any of the dimensions and environmental conditions fall outside the total range for a specific period, or comprise a mix of values that fall within the total range, but across a range of periods, proceed to Scenario 4.

Scenario 1: Bats, droppings, and/or bat-flies

If bats, droppings and/or bat-flies are recorded, the PRF is a roost and a European Protected Species Licence (EPSL) will therefore be required if there is any potential that the roost might be destroyed or damaged, or bats disturbed as result of any proposed action. Where the evidence comprises droppings or bat-flies alone, or a combination of the two, careful interpretation of the data gathered at each visit will be required in order to provide a reasoned prediction of the species (unless conclusively identified by DNA), the number of bats and the purpose the PRF serves (i.e. maternity, mating, winter, transitory, etc.). Note: when considering the potential for damage and disturbance, it is sensible to consider not only direct impacts, but also indirect impacts (see the Indirect Damage and Disturbance, and Comparative Assessment of Environment discussion in Chapter 12).

Scenario 2: Associated dimensions and environment – *"reasonable likelihood"*

If no conclusive evidence is recorded but the dimensions and environmental conditions are *all* within the range of *"reasonable likelihood"* for a specific bat species within a particular period of the year, and the internal volume is sufficient, it may be concluded that there is a *"reasonable likelihood"* the PRF is exploited as a roost by that species in that period. This conclusion will be strengthened if:

» An associated competitor has been recorded occupying the PRF.

» The bat species has been recorded during any other presence/absence surveillance (such as ultrasound or trapping).

However, without conclusive evidence, an application for an EPSL may not be appropriate. Nevertheless, it might be sensible to set out the evidence recorded with an appropriate impact assessment that assumes the PRF is a roost, alongside a due-diligence safeguarding strategy (such as that set out in the section dealing with Inconclusive and Null Results in Chapter 12), and seek advice from the Discretionary Advice Service offered by the appropriate government agency.

Scenario 3: Associated dimensions and environment – Potential

If the dimensions and environmental conditions fall within either a mix of *reasonable likelihood"* and the total range, or solely within the total range, for a specific bat species within a particular period, and the internal volume is sufficient, then it may be concluded there is 'potential' that the PRF is suitable and might be a roost. As with comprehensive *"reasonable likelihood"*, this conclusion will be strengthened if:

» An associated competitor has been recoded occupying the PRF.

» If the bat species has been recorded during any other presence/absence surveillance (such as ultrasound or trapping).

Despite the perceived low level of risk, if there is any potential for damage to the PRF, it may be sensible to offer a generic due-diligence safeguarding strategy (such as that set out in the section dealing with Inconclusive and Null Results in Chapter 12).

Scenario 4: Null result

If any of the dimensions and environmental conditions fall outside the total range for a specific period, or comprise a mix of values that fall within the total range, but across a range of periods, then a reasoned argument will be required if the feature is to continue to be regarded as a PRF. Notwithstanding, in any situation where there is an interval between a null surveillance result and operations that will result in the loss or damage to a tree or larger area of wooded habitat, it is sensible to take a cautious approach and offer a due-diligence safeguarding strategy. The reasons for this, along with a generic strategy, are provided in the section dealing with Inconclusive and Null Results in Chapter 12.

CHAPTER 9

Remote-Observation

In this chapter	
Introduction	The data the method is to collect
Confounding factors	Bat ecology, human biology, wooded environments and the solution
Equipment	The equipment needed to collect the data
Health and Safety	Considerations in respect of biotic and fatigue hazards
Remote-observation method	How to collect and record the data
Interpretation	How to present and interpret the data

9.1 Introduction

Remote-observation comprises a visual-observation from a static-viewpoint held over a predetermined period of time to see whether bats or any competitor/predator emerge, return, visit or investigate the PRF.

9.2 Confounding factors

9.2.1 Bat ecology

The first consideration when embarking on a remote-observation survey is the behaviour of the bats themselves. Key considerations that apply to all situations involving wooded habitat and individual trees comprise the following:

» The times the individual species typically emerge from, and return to, roosts.

» The behaviour of bats generally upon emerging.

» The behaviour of bats generally upon returning.

» The number of bats present.

» The situation.

Emergence and return behaviour

To perform robust remote-observation surveillance the surveyor must be able to maintain visual contact with the PRF for the entirety of the observation period. It is therefore necessary to understand the temporal range across which each bat species can be predicted to emerge and return in each period.

In order to define the temporal 'window' within which individual species emerged and returned, Andrews and Pearson (2016) reviewed empirical data reported for bat species occurring in the UK. The review specifically searched for the standard deviation[1] in respect of average emergence and return times, because an SD of 1 illustrates the interval in which 68% of emergence and return times occurred, and might therefore be rationally applied in the context of *"reasonable likelihood"*. Unfortunately, the review found that reporting between studies was not identical in scope, and in many papers the information comprised only one or a combination of: mean average; median, standard deviation; and range (i.e. the full span between the earliest and latest emergence and return).

Table 9.1 provides a summary of the review. Where times are reported for both females and males, these are provided. Where a single SD figure is reported, this is adopted. Where more than one SD is reported, the outlier times are adopted. Where SD was not reported, the range was adopted in preference to the mean. It is recommended that the full review be consulted (see www.battreehabitatkey.com) and the individual studies individually referenced in any subsequent report.

> The standard deviation surveillance window represents a balance between confidence and proportionality in the context of an Impact Assessment in support of planning. However, as the studies that investigated emergence and return behaviour will in most cases have been recorded in locations far removed from the surveillance location (even in another country), and also in situations that may be significantly different from that present in the surveillance site (even house roosts rather than trees), the rarity of the bat species potentially present should be taken into account. In some situations, it may be wiser to consider the full range of times across which emergence/returns occurred and extend the observation period accordingly.

Behaviour upon emerging

No bat species needs to drop to gain sufficient speed to gain altitude; even noctules can exit the roost on an even plane and at an upward angle. Natterer's bats have been observed emerging from a roost and climbing round a limb to a 'launch' position away from the roost entrance. Natterer's bats, noctules and brown long-eared bats have emerged from roosts without registering on ultrasound detectors.

It is worth stressing that bats do not emerge as a swarm like bees, or like shot from a gun, nor do they emerge at regular intervals like parachutists from an aircraft, but as ones and twos at irregular intervals. In addition, the bat(s) may not exit on a flight-path toward the surveyor, but may turn immediately upon emerging and head in the opposite direction.

Behaviour upon returning

Groups of some species may swarm outside tree-roosts prior to entering at dawn. This is particularly common with Daubenton's bats, which not infrequently swarm in sufficiently high light-levels to allow the phenomenon to be seen clearly without the use of night-vision. Barbastelle groups and Natterer's bats occasionally swarm but almost always in total darkness. However, this behaviour is far less typical for the other species, which

---

1   Standard deviation (SD) quantifies the variation on either side of the mean in which proportions of data occurred. If the data are 'normally distributed', 68% will fall within 1 SD, 95% within 2 SDs and 99% within 3 SDs.

Table 9.1 Summary review of the emergence and return times reported for tree-rooting bat species occurring in the UK

Species	Reported emergence times in relation to sunset	Reported return times in relation to sunrise
Barbastelle	SD 17–31 minutes after	SD 4 hours and 14 minutes to 2 hours and 15 minutes before
Bechstein's bat	R ♀ 2 minutes before to 1 hour and 32 minutes after R ♂ 25 minutes before to 1 hour and 27 minutes after	R ♀ 1 hour and 17 minutes to 9 minutes before R ♂ 6 hours and 53 minutes before to 2 minutes after
Alcathoe's bat	SD 16 minutes before to 14 minutes after	R 33 minutes before to 17 minutes after
Brandt's bat	SD 18–20 minutes after	SD 35–16 minutes before
Daubenton's bat	SD ♀ 36–56 minutes after SD ♂ 22 minutes to 1 hour and 25 minutes after	R ♀ 7 hours and 30 minutes to 10 minutes before SD ♂ 4 hours and 59 minutes to 5 minutes before
Natterer's bat	SD 54–57 minutes after	SD 50–30 minutes before
Whiskered bat	M 33 minutes after	M 2 hours and 7 minutes before
Leisler's bat	SD 8–27 minutes after	M* 5 hours and 20 minutes to 12 minutes before
Noctule	SD 16 minutes before to 31 minutes after	R Onset of civil twilight to 3 minutes before sunrise
Nathusius' pipistrelle	R 11–50 minutes after	R 60 minutes before to sunrise
Common pipistrelle	SD 6–43 minutes after	SD 4 hours and 50 minutes before to 1 hour and 6 minutes before
Soprano pipistrelle	SD 12–55 minutes after	SD 6 hours and 18 minutes before to 2 hours and 40 minutes before
Brown long-eared bat	SD 28 minutes to 1 hour and 34 minutes after	SD 1 hour and 31 minutes to 1 hour and 13 minutes before
Lesser horseshoe-bat	SD 30–36 minutes after	SD 41–19 minutes before

SD, standard deviation; R, range; M, mean.

* No SD or range given, so timings taken from two reported means.

appear to swarm even less when occupying trees than they might do when occupying houses.[2]

Swarming aside, the interval between the first bat emerging after sunset is typically significantly shorter than the interval between the final bat returning and sunrise. As a result, the first bat emerging in the evening may do so in higher light-levels than the last bat returning in the morning. The same is also true of the last bat emerging in the evening and the first bat returning at dawn.

The surveillance team should also bear in mind that the emergence times given in Table 9.1 are not the times the first and last bat emerged and returned. The times represent the times the first bat emerged in relation to sunset over multiple observations; *the times last bat emerged are not reported but are likely to be significantly later*. In the same way, the times given in respect of returns, are the time the last bat returned over multiple observations; *the times the first bat returned are not reported, but are likely to have been significantly earlier*.

Even taking the above into account, looking at the wider range of SD and range in respect of reported emergence and return times, it can rationally be accepted that the duration of emergence may be shorter (i.e. the interval between the first and last bat emerging) than the duration of the return (i.e. the interval between the first and last bat returning). It is therefore rational to suppose that a dawn-return survey will need to maintain concentrated focus for a significantly longer period than an evening-emergence survey if the results are to be considered reliable, and if an accurate count is to be achieved.

The number of bats

The number of bats present also influences emergence and return behaviour in terms of:

» Whether emergence will occur at all – individuals may not emerge every night, even in superficially optimal conditions.

» The time emergence and return may commence – individual males tend to emerge later than females and, having made one foraging bout, return to the roost without emerging again that night.

» In the case of groups, BTHK project surveillance has not registered the so-called 'light-sampling' behaviour in respect of trees, but has perceived what was taken to be encouragement behaviour in late-summer, where mothers had newly volant young that appeared reluctant to leave the roost. In the case of individual bats, all emergence witnessed has occurred with the bat immediately dispersing to forage. Similarly, returns have occurred without any swarming behaviour observed, and at times at such high speeds as to barely register on video-footage.

---

2 Questions that will require answering before swarming can be factored into any interpretation of the results of a return observation would include: (i) Which species swarm and which do not? (ii) Is there any geographic variation in swarming behaviour? (iii) Is there any situation variation in swarming behaviour (i.e. is there any association between swarming and the position of the tree in the habitat, or the PRF on the tree)? (iv) Is there any sexual variation (i.e. do all-male groups swarm, or all-female, or both)? (v) Is there any seasonal variation? (vi) Does swarming comprise all the bats that will occupy that roost on that day, or are some of the group already in occupation? (vii) When does it begin in relation to sunrise, and how long does it last (i.e. what is the interval a transect might cover in order to assess the status of multiple PRFs)?

9.2.2 Human biology

Having reviewed emergence and return behaviour, it is readily apparent that the greater proportion of tree-roosting bats emerge and return in full darkness. It is therefore necessary to assess the ability of the survey team to see in low light levels.

Modes of vision

In diminishing levels of light, the human eye progresses through three vision modes over a period of up to 45 minutes. These vision modes comprise:

» **Photopic-vision:** The operational mode in daylight and artificially illuminated conditions, and characterised by:

– Sharp focus; and

– Colour vision.

» **Mesophic-vision:** The operational mode 15 minutes after sunset, and characterised by focus being unbalanced between central and peripheral vision with a trade-off between sensitivity at the expense of resolution. This switch results in the best field of focus being slightly off to the side. This mode is the least effective for focused observation, but due to the high level of ambient light in our night sky (particularly in overcast conditions), it is typically where our vision rests.

» **Scoptic-vision:** The operation mode in a cloudless moonless night (even moonlight may prevent adjustment to scoptic-vision), and characterised by:

Poor focus; and

– No colour vision (Green 2013).

As the emergence may take anything up to and even over an hour, the light-levels pass through different phases which the US Naval Observatory has defined as:

» **Civil Twilight:** The center of the sun is geometrically 6° below the horizon. At the onset of Civil Twilight lux-levels in open habitat are around 10.75 lux with sufficient illumination (under good weather conditions) for terrestrial objects to be clearly distinguished. However, by the end of Civil Twilight illumination has reduced to 3.4 lux (Martin 1990). During Civil Twilight the human eye passes from photopic-vision and into mesophic-vision.

» **Nautical Twilight:** The centre of the sun is geometrically 12° below the horizon. Lux-levels in open habitat are around 1.08 lux with illumination sufficient for general outlines of ground objects to be distinguishable, but detailed outdoor operations not possible, and the horizon is indistinct. At Nautical Twilight the human eye remains within mesophic-vision.

» **Astronomical Twilight:** The centre of the sun is geometrically 18° below the horizon. Depending upon the phase of the moon and the level of cloud, the human eye either remains in mesophic-vision, or enters scoptic-vision. Examples of lux-levels in cloudless situations are as follows:

– Full moon – c. 0.108;

– Quarter moon – c. 0.0108;

– Starlight – c. 0.0011; and

– Overcast sky *or under the woodland canopy* – c. 0.0001.

The ability of humans to see effectively is impeded at levels below 10 lux. At light-levels of 3.4 lux, visual acuity is reduced by c. 50% (Lewis and Taylor 1964). Nevertheless, the human eye can still technically 'see' landscape elements (i.e. differentiate between a tree and a hedge) down to the 0.108 lux provided by a full moon, but below this we are at a significant disadvantage, even to the point of a risk of collision with static objects.[3] However, even when Astronomical Twilight comprises zero-cloud and a full moon, the aspect of the light may in fact work against us depending upon the environment.

9.2.3 Wooded environments

In a building situation, the survey is typically performed in an open environment with an unobstructed view of the bat, which descends to take advantage of cover beneath the roost entrance.

In a tree situation, it is often difficult to get an unobstructed view of the roost entrance without either being at an acute angle or a significant distance away. Furthermore, to get that view the surveyor is typically looking through a 'window' in the foliage, and that window may move with the wind.

Finally, the bats may emerge and disappear immediately into the canopy, or into the shrub-layer which will mean there is an obstruction between the surveyor and the bat which may comprise several inches of wood. Individual bats may also choose different flight-lines as they exit and a pair may emerge together, but one may go up, another down and then suddenly a third exit on an even plane, and all at different speeds.

Even where scoptic-vision is achieved, the poor focus resolution and speed works against any surveyor attempting to detect a moving target. Furthermore, fixed focus upon a PRF entrance may alter our perception due to local retinal adaptation; staring at a black shape in a lighter background will result in a reversal image imprint when the surveyor looks away from the PRF (Green 2013).

If this were not bad enough, the aspect of both natural and artificial illumination also plays a significant role in our ability to focus.

During the day we tend not to notice the aspect of the sun, because for the greater part it is above the tree-line and offset to our line of sight. In the evening, we notice not only the aspect of the sun, but also the angle of the moon, because both are below the tree-line and within our line of sight. This may have a significant detrimental effect upon our ability to see a PRF. To illustrate, on a clear moonlit night, if the moon was directly behind a surveyor it might assist the survey providing the air-space between the moon and the PRF entrance was unobstructed. However, if there was any canopy between the surveyor and the PRF, the shading effect would darken the PRF entrance, and the foliage would be illuminated and therefore the surveyor's eye would remain in mesophic-vision and attention would be drawn to the foliage rather than the PRF. If the moon was at any aspect other than immediately behind the surveyor, regardless of shading, the effect would be to place the PRF in shadow and the light in the surveyor's eyes.

At dawn, the situation is reversed and it is the sun that decides how visible the PRF entrance may be. However, it should be borne in mind that even mesophic-vision will only be achieved after a period of adjustment.

9.2.4 The solution

If we accept that a professional survey should be sufficiently detailed to confidently detect all bat species, in order to establish whether there was *a "reasonable likelihood"* that a specific PRF was or was not being exploited by roosting bats on the day the surveillance took place, the surveillance team should logically have the facility to maintain the observation at the same level of clarity across a wide range of distances, and over the full

---

3   It is an irony that having completed an evening-emergence survey that has attempted to detect a tiny flying object with no optical assistance whatsoever, the surveyor will use a torch to find his/her car …

surveillance period, regardless of the light-level. Furthermore, if the surveillance is being performed for a third party, the team should have the facility to record the observations in order that the results are available for review.

Finally, the surveillance will be improved if it has the facility to identify any other the organism that is exploiting the PRF, as this will allow the results to be compared with the Database in order to search for associations that might hint at the suitability of the PRF even in the absence of the bats themselves.

The use of a zero-lux camera and an infrared lamp or an equivalent night-vision device immediately removes all the constraints in terms of the behaviour of the bats, the weakness of human night-vision, and the environment for a wide range of reasons, including:

» The camera is not subject to gradual adaptation in line with the ambient light levels.
» Where the eye adapts locally, the camera sets the exposure globally; it is not therefore subject to imprinting and sees both entrance and background (and a sphere around both).
» The camera does not get tired and lose concentration, or blink, or sneeze, or need to go to the toilet.
» It can focus-in to give sufficient detail that it can detect the exit and entrance of even invertebrate competitors.

Finally, a zero-lux camera will provide a conclusive and comprehensive record of the observation, both positive and negative, if the result of surveillance is later challenged.

The evidence presented demonstrates unequivocally that:

» Evening-emergence is more cost-effective in respect of the duration of the observation; **and**
» Only with the use of both night-vision equipment and an ultrasound detector is there a *"reasonable likelihood"* that if bats are present they will be detected with sufficient information to determine the time of emergence, and the species and numbers present.

> **Note:** Failure to take account of this information is to wilfully ignore evidence in respect of bat ecology, human biology and wooded environments.

9.3 Equipment

For remote-observation to be fully effective and the results reliable, it will require:

» A ground-truthing account.
» A recording form.
» Zero-lux video camera with an infrared lamp or array of lamps, or an equivalent night-vision device sufficient to illuminate the full extent of the PRF entrance and at least 0.5 m on either side with a means of recording footage in real time.
» An automated ultrasound 'bat-detector' that can be left unattended to record in full-spectrum.
» A means of monitoring the temperature.
» A GPS.
» Some means of telling the time.

The use of camera-traps to detect roosting bats has been found to be effective in some situations. Positive encounters have been made with individual bats investigating PRFs in a sequence of approaches, also females swarming round roosts upon returning to suckle dependent young, and groups swarming prior to entering a roost at dawn. However, these units have not been proven to be *consistently* reliable for detecting emergence of groups or individuals, or individuals returning to roost alone, or colonies returning at widely spaced intervals without swarming in the proximity of the roost entrance. Furthermore, even where bats have been recorded, the quality of the footage achieved has typically been insufficient to identify them to a species.

At present, all models have either timed short-duration video-capture, or are triggered by a passive infrared (PIR) sensor. This suggests a trail-camera would need to be deployed relatively close to the PRF for the PIR to register an emerging bat, but the closer the unit was deployed to the entrance, the narrower the field of vision would be, and therefore the shorter the amount of time the bat was in view of the camera. As any delay in activation means that the bat is in flight in the footage recorded, unless the footage captures the bat immediately returning to the entrance, it is speculative to conclude the bat emerged from the target PRF. In the same way, the sensitivity of the PIR and the camera quality mean that the camera has to be significantly high to allow a trail camera to be deployed sufficiently far from a PRF for it to detect a bat early enough to have time to activate the video and thereby capture an individual bat returning if that bat heads straight to the roost entrance and does not remain on the wing in the vicinity of the PRF.

It is hoped that in time a camera-trap will be developed that is small enough to be deployed at height, but has sufficient memory and power-supply to facilitate timed activation, long-duration capture (ideally 3 hours), timed deactivation and long-term deployment (ideally 57 days, but 31 would be a useful start). If a PIR were also included to put a time-stamp on any movement registered, this would represent a significant step forward in tree-roost detection, particularly in light of the need for sequential surveillance observations (see Chapter 11).

» Some means of telling what the temperature was at the start and end of the observation period.
» A deck-chair.
» A head-torch.

The ground-truthing account
This will have been created at the ground-truthing stage (see Chapter 6).

The recording form
The recording criteria adopted for the BTHK project and set out at Table 9.2 were defined specifically to focus attention to the meaningful data required to fulfil the objective (a Word copy of this table can be downloaded from www.battreehabitatkey.com). When submitting a record to the Database all the attributes should be assigned a value.

Zero-lux camcorder
Zero-lux camcorders have existed for well over a decade. The Sony DCR-SR35E was released in 2008 but is still favoured amongst bat surveyors across the world. As a result of its relative age, Sony DCR-SR35Es frequently appear on eBay and Amazon for £100 or thereabouts. However, a memory card and a high-capacity battery are required to operate the unit for collection of prolonged footage (in the case of the DCR-SR35E, at the time of writing a Duracell DR9700B typically costs in the region of £30). These cameras are perfectly acceptable for remote-observation. However, if 'television quality' is required, the Canon XA10 will set you back between £1,100 for the XA10 and £1,800 for the latest model (currently the XA35).

Table 9.2 The BTHK remote-observation recording form

SURVEYOR NAME(S)		
DATE		
TREE LOCATION	OS Grid ref:	
CAMERA LOCATION	OS Grid ref:	
TIMING		
SURVEY COMMENCED		
SUNSET TIME		
SURVEY COMPLETE		
CONDITIONS UPON COMMENCING		
CLOUD-COVER		
WIND		
RAIN		
TEMPERATURE		°C
CONDITIONS UPON COMPLETION		
CLOUD-COVER		
WIND		
RAIN		
TEMPERATURE		°C
RESULT		
DROPPINGS		
EMERGENCE OR RETURN?		
FAMILY/SPECIES		
COUNT		
VISITATION		
FAMILY/SPECIES & NUMBER		
INVESTIGATION		
FAMILY/SPECIES & NUMBER		

Regardless of the camera, however, the most important piece of kit is the infrared lamp. In most situations, infrared floodlights such as those available for security situations typically lack the range and intensity required for remote-observation of small targets in woodland.

The best units are hunting lamps such as the Nightmaster 800-IR (starting price c. £160) that will illuminate a PRF from up to 100 m, away allowing the zero-lux camera to zoom in without the bats being aware at all. An alternative is the Evolva range of torches available from Amazon: the T20, T38 and T50 units retail in the region of £30–40 but require the additional purchase of high-capacity batteries and charger. Regardless of the lamp, a sturdy clamp will be required to fit it to the top of a tripod, and the Manfrotto 'superclamp' is an excellent choice at c. £20–30.

A lighting and camera configuration using a single tripod, a 'hot-shoe bracket' and a 'tripod ball-head' is shown in Figure 9.1. This equipment is carried as standard by the BTHK project and allows a climbing team to quickly and easily deploy a zero-lux camera

Figure 9.1 A lighting and camera configuration using a single tripod, a 'hot-shoe bracket', a 'tripod ball-head', a Manfrotto superclamp and both a Sony DCR-SR35E (equipped with a Duracell DR9700B) and an Evolva IR torch. All are available from Amazon.

in any situation where bats have been encountered during a close-inspection but where an identification and/or a count could not be ascertained.

With this equipment, the team zoom in with the camera and capture the entire session for review later if necessary. With video, you have proof that the roost is there and something tangible to share with your client. Video is also very easy to review, and most surveyors can reliably review up to four films playing simultaneously on individual screens. Video also puts a little interest back into the method and, where bats and/or competitors are encountered, rewards the survey team with a deeper insight into roosting ecology.

Hand-held thermal imaging cameras are still relatively new technology, and prices are high when compared to zero-lux cameras. The BTHK project has trialled different units, but found that although all would certainly detect bats emerging, the resolution in terms of identification was unsatisfactory both for the bats themselves and invertebrate competitors. Although the units may have value for detecting whether or not a PRF is occupied by a group of bats or a competitor such as a bird or squirrel, this is all the resolution that can be achieved without a close-inspection (i.e. the conclusion of a thermal imaging inspection is that there appears to be 'something' occupying the PRF, but what it is we do not know) and where individual bats are roosting well away from the roost entrance, the risk of a false-negative is significant.

Automated ultrasound detector
If reliable identification is even to be attempted for the *Myotis* bats, a full-spectrum detector will be required. However, it should be borne in mind that in woodland, a detector may have a maximum effective range of 10 m, assuming that the bats do actually echolocate as they leave/return.

GPS
Individual trees in woodland should be assigned a 'waypoint' on a GPS if the tree is to be found in the dark for a dawn-return observation (although taking into account the information in Table 9.1, it might be more sensible to stay out all night). In addition, the location where the camera was deployed should be accurately plotted so the account of the surveillance method is complete.

Clock
A means of telling the time will be required to ensure the observation does in fact begin early enough and conclude late enough.

Thermometer
A thermometer is required to identify whether the temperature over the full range of the surveillance period was within that above which bats typically forage: 7°C (see 'Assessing the conditions' at subsection 9.6.2).

A deck-chair
Remote-observation has the potential to be mind-numbingly dull. It is also uncomfortable to stand looking up at a single point for hours on end, and this uses up energy. It is far more sensible to accept this and carry a deck-chair to ensure energy is conserved and the surveyor remains alert. This is because the camera may need refocusing from time to time, and the tree may even begin to move in a breeze and the focus may need to be brought out to a wider field of vision and the IR lamp repositioned or the diffuser adjusted.

If you wish to perform the survey in total comfort, a pop-up chair-hide protects the equipment and the surveyor and comes in a bag that can readily be carried on the surveyor's back with the necessary recording form, etc., leaving the hands free to carry the camera and lamp already fitted to their tripods. A 'Nitehawk' hide is illustrated in Figure 9.2 and may be purchased from Amazon.

Figure 9.2 A 'Nitehawk' pop-up chair hide. Available from www.amazon.co.uk.

A head-torch
A head-torch will be required to operate equipment, decommission and pack away, and navigate in the dark. The torch should have a red filter for use during the observation.

9.4 Health and Safety
It is up to the individual team to ensure they have adequate insurance and a risk assessment. However, it cannot be overstressed that insect-repellent should be liberally applied and the survey team should ensure that they have sufficient rest before and after an observation for them to be safe to drive and within the legal limits with respect to work-breaks, in accordance with the current legislation.

9.5 Remote-observation method
Remote-observations are not simply about looking for bats, remote-observations are about searching for all the clues that might hint that the PRF is exploited as a roost.

A thorough observation will comprise five stages:

» **Stage 1** – A review.
» **Stage 2** – An assessment of conditions.
» **Stage 3** – A search for droppings.
» **Stage 4** – Setting-up.
» **Stage 5** – The observation.

By completing all five stages carefully, sufficient data will be provided for a meaningful analysis.

Constraints

A full list of constraints is provided in Chapter 7, but in summary the primary constraints associated with the efficacy of remoteobservation are the following:

» Unless the observation is being performed to gather missing count or identification data following a close-inspection, it may be unknown whether the PRF is in fact suitable prior to the survey commencing.
» The method cannot detect droppings inside the PRF or changes to the internal environment.
» Not all bats observed can be identified to species.
» Even where a group is observed, the sex of some species cannot be inferred.

9.5.1 Remote-observation: Stage 1 – Review

The first stage of every observation should be the review of the results of the ground-truthing, and any pre-existing account(s) of earlier observation(s).

In addition, reference to the accounts of competitor species associated with the presence of specific bat species (provided in Chapter 8) will identify the competitors that exploit the same environment. This will widen the objective of the observation to recording organisms that occur more frequently than the bats themselves, which is likely to result in an increased level of focus during prolonged observation.

As accounts of the observations build, these should be collated into a format that can be taken into the field by the surveillance team. Thereafter, this review can be compared against the situation encountered during later observations to identify any changes in the presence of bat or competitor.

It cannot be overemphasised: remote-observations should aim to identify any changes in the occupancy of the PRF by other organisms. If the accounts of previous observations are not collated and reviewed by the survey team immediately prior to each subsequent observation, the surveillance will lack a key element of focus. It is far better for a surveyor to be thinking *'if nothing has changed, I should be seeing slugs emerge'*, than to have no advance intelligence. The statement *'no bats emerged/returned'* gives little insight. The statement *'no bats emerged, but woodlice/slugs/blue tits/grey squirrels were observed entering and leaving'*, gives far greater confidence that nothing has been missed. Conversely, any change (no matter how minor) will immediately heighten attention and ensure the observation is thorough. **No one performs remote-observation surveillance in the hope that they won't find bat roost!**

9.5.2 Remote-observation: Stage 2 – Assessing the conditions

Before any observation is considered, the weather conditions must be very carefully assessed to ensure they will remain suitable for the full duration. The conditions that must be considered, and their order of importance, are:

1. Temperature.
2. Wind.
3. Rain.

Temperature

Providing the temperature is above 7°C and the wind speed is below 2 miles per hour (mph), in most cases it is light intensity that dictates the time of invertebrate flight; temperatures above this threshold only influence altitude (Lewis and Taylor 1964).

Billington (2002) found that a radiotagged barbastelle *Barbastella barbastellus* did not emerge on a night where the temperature had dropped to 0.6°C despite emerging on the night before where the temperature was 14°C and emerging the night following where the temperature was 7°C. A similar situation was reported for Leisler's bats, with activity ceasing altogether when temperatures fell below 6°C (Hopkirk and Russ 2004).

However, as different bat species feed on different invertebrates using different hunting strategies, some species may continue foraging below the 7°C threshold. For example, brown long-eared bats favour gleaning as a foraging strategy and have been demonstrated to forage in temperatures as low as 3.5°C (Entwistle *et al.* 1996), and a study of Bechstein's bat recorded three females of the species emerging from different roosts to forage on an evening with a starting temperature of 7°C after which it dropped to 1°C (Palmer *et al.* 2013). In contrast, most species of moth only take wing at temperatures above 12°C (Ransome and Hutson 2000) and those species that target moths in flight will be likely to favour emergence on warmer nights.

Regardless, even if activity does not cease altogether, bat activity is significantly reduced in temperatures below 7°C (Avery 1985; Linton 2009) and this must be taken into account if surveillance results are not to be considered inconclusive. This is particularly important for return-surveys, where the temperature has fallen below an acceptable threshold for the target species following emergence, as they may return earlier than anticipated.

Therefore, *for an observation to be valid, as a minimum the temperature should still be above the minimum threshold at the end of an evening-emergence survey, and at the start of a dawn-return survey.*

The BTHK Project commissioned the Met Office to analyse their temperature data in order to define the temperature that a surveillance observation would have to commence, half an hour before sunset, for it to be 'more-likely-than-not' that the observation would conclude two hours following sunset at a temperature that was above the 7°C threshold. Funds being limited, a single central location was chosen: Birmingham. The analysis was run in all 12 months of the year with data spanning 55 years (1961–2016). The results are set out as percentage probabilities in Table 9.3.

The data set out in Table 9.3 demonstrate that it is not always the case that autumn-flux and winter months are too cool for bats to take wing in the early part of the night, and it is not always so warm that bats will emerge in the spring-flux, pregnancy, nursery and mating periods;[4] care must be taken to check the forecast prior to commencing an observation.

Wind

Referring back to the previous section, most flying invertebrates favour wind speeds of 2 mph or less (Lewis and Taylor 1964). As a result, as wind speeds increase, flying invertebrates seek shelter and remain closer to the ground and nearer to vegetation that offers

---

4 See also the discussion regarding activity in Chapter 11.

Table 9.3 The % probability that a surveillance observation that commenced at a specific temperature 30 minutes before sunset would conclude two hours later at a temperature that was 7°C or above, shown for all months of the year

Temp. °C	% Probability											
	Winter		Spring-flux		Pregnancy		Nursery		Mating		Autumn flux	
	Jan	Feb	Mar	Apr	May	Jun	Jul	Aug	Sep	Oct	Nov	Dec
31.1–32	—	—	—	—	—	—	—	100	—	—	—	—
30.1–31	—	—	—	—	—	—	—	100	—	—	—	—
29.1–30	—	—	—	—	—	—	—	100	—	—	—	—
28.1–29	—	—	—	—	—	—	—	100	—	—	—	—
27.1–28	—	—	—	—	—	—	100	100	—	—	—	—
26.1–27	—	—	—	—	—	100	100	100	100	—	—	—
25.1–26	—	—	—	—	—	100	100	100	100	100	—	—
24.1–25	—	—	—	—	—	100	100	100	100	100	—	—
23.1–24	—	—	—	—	100	100	100	100	100	100	—	—
22.1–23	—	—	—	—	100	100	100	100	100	100	—	—
21.1–22	—	—	—	—	100	100	100	100	100	100	—	—
20.1–21	—	—	—	—	100	100	100	100	100	100	—	—
19.1–20	—	—	100	100	100	100	100	100	100	100	—	—
18.1–19	—	—	100	100	100	100	100	100	100	100	—	—
17.1–18	—	—	100	100	100	100	100	100	100	100	—	—
16.1–17	—	—	100	100	100	100	100	100	100	100	100	—
15.1–16	—	—	100	96.9*	100	100	100	100	99.5	98.1	100	—
14.1–15	—	—	100	100	99.3	100	100	100	99.2	99.3	100	100
13.1–14	100	100	91.9	97	98.5	99.5	98.8	100	99.1	95.9	100	100
12.1–13	100	100	92.3	95.7	95.9	98.9	98.7	98.1	97.6	97.2	98.9	97.6
11.1–12	100	94.6	88.9	90.3	89.3	96.3	100*	94.4	92.3	92.8	96.5	98.5
10.1–11	94.9	88.9	84.5	77.1	83.7	90.9	87.5	100*	91.7	87.3	92.8	95.6
9.1–10	87.9	78.4	72.9	55.9	71.3	70	—	—	76	74.9	86.1	85
8.1–9	69.9	67.5	45.5	36.5	49	70.6	—	—	90	46.8	58.4	70
7.1–8	36.4	29.1	15.8	12	17.2	20	—	—	25	23.4	33.8	25.4
6.1–7	10	2.9	2.9	1.9	0	0	—	—	0	5.4	10.3	8.4
5.1–6	1	0.0	1.1	0	0	0	—	—	—	12.5	3.5	2.8
4.1–5	2.6	1.2	0	0	0	—	—	—	—	0	0	0.6
3.1–4	1.3	1.5	0	0	0	—	—	—	—	0	1.2	1.2
2.1–3	0	0.8	0	0	0	—	—	—	—	0	1.5	0
1.1–2	1.8	0	0	0	0	—	—	—	—	—	0	0
0.1–1	0.9	0	0	0	—	—	—	—	—	—	0	0

Source: Met Office data.

* Anomalies caused by atypical warm or cool weather fronts passing.

a vertical shield. Many tree-roosting bat species appear relatively tolerant of light breezes, but this is not true for all species. For example, Linton (2009) found that the activity of noctules was considerably reduced at wind speeds of over 3.4 mph (1.5 m/s).

Wind is recorded using the Beaufort Force Scale. For ease of reference the scale is provided below and colour-coded **green** for suitable conditions, amber for poor conditions and **red** for unsuitable conditions:

0   CALM: Wind 0–1 mph – Smoke rises vertically as a ribbon and dissipates some distance from the source.

1   LIGHT AIR: Wind 2–4 mph – Smoke drifts and indicates a wind direction, but foliage is stationary.

2   LIGHT BREEZE: Wind 5–8 mph – Wind is now perceptible on exposed skin and foliage is rustling.

3   GENTLE BREEZE: Wind 9–12 mph – Foliage and small twigs are constantly moving.

4   MODERATE BREEZE: Wind 13–18 mph – Small branches are moving.

5   FRESH BREEZE: Wind 19–24 mph – Branches of a moderate size are moving. Small trees in leaf are swaying.

6   STRONG BREEZE: Wind 25–31 mph – Large branches are in motion.

7   NEAR GALE: Wind 32–38 mph – Whole trees are in motion.

8   GALE: Wind 39–46 mph – Some twigs are broken from trees.

9   STRONG GALE: Wind 47–54 mph – Branches may break off trees. Small trees may be blown over.

10 WHOLE GALE: Wind 55–63 mph – Larger trees may be broken off or uprooted.

11 STORM: Wind 64–72 mph – Widespread vegetation and structural damage is likely.

12 HURRICANE: Wind 73+ mph – Severe widespread damage to vegetation and structures is likely.

Regardless of the habitat or where in the world the surveillance is being performed, Force 0 and 1 are the typical conditions under which a remote-observation survey commences, but these frequently give way to Forces 2 and 3 as the sun sets. This is due to the rapid changes in the ground temperature, which influences wind speed and direction. No site is exempt from this.

No observation would sensibly commence in conditions above a light breeze, and the results of any observation that completed in a Force 4 moderate breeze should be treated with caution. A fresh breeze or above represents unsuitable conditions (for bats and surveyors if they are under the canopy of a tree!).

Where the target of remote-observation is a structure, the PRF entrance will not move. Where the target of remote-observation is a tree, the PRF may very well move, and if it is over 6 m up that movement may be significant even in a light breeze due to the sail effect of the canopy. Breezes are not constant, but are interspersed with calms: as a consequence the cumulative effect is that the entire tree begins to sway. Although this may only be gentle, the movement may move a close-focused PRF out of focus, and even out of shot altogether!

Therefore, cameras should be monitored constantly throughout the observation period to ensure no drop-out in focus.

Rain

Rain has been shown to prevent emergence in brown long-eared bats (Entwistle *et al.* 1996) and although in Ireland colonies of Leisler's bats tolerated light rain, they either did not emerge at all under heavy rain or returned to the roost within minutes of having departed (Shiel and Fairley 1999).

Whilst it is up to the individual surveillance team to judge the conditions, it is common sense that commencing or continuing an observation in anything more than the lightest rain would be unwise, if for no other reason than the fact that unless the equipment is sheltered, even unexpected light showers of short duration may have catastrophic effects upon the video camera and any associated lighting.

9.5.3 Remote-observation: Stage 3 – Search for droppings

Even though the PRF may not be accessible, the first stage of the remote-observation should still be to perform a thorough search for droppings on the ground beneath the PRF, and on any field- and shrub-layer foliage further out from the entrance (see Chapter 8).

9.5.4 Remote-observation: Stage 4 – Set-up

Stage 4 should proceed as follows:

1. Arrive at the site sufficiently early that having walked to the target tree, the surveyor will have no less than 1 hour to set-up and then acclimatise to the situation and take in his/her surroundings prior to the time the first bat might emerge.

2. Take into account that until the light-levels drop sufficiently, it may not be possible to switch the camcorder to night-shot mode or focus the infrared lamp accurately upon the PRF – continuous adjustments may have to be made as the observation progresses.

3. Take into account the fact that the PRF may begin to move and, where possible, choose an observation position where the camera and lamp can be adjusted to a wider field of vision without having to redeploy them in another location. If, however, there is no other way of achieving coverage, try to identify a fallback position prior to the commencement of the survey.

4. Take into account that bats roosting in trees are not typically acclimatised to noise outside the roost. Experiences hint that bats that have been disturbed with a close-inspection earlier the same day may begin to emerge earlier than they might have otherwise. In the case of barbastelles, if you cough near the roost they may very well decamp; when it is a dead tree and you are looking at a bark plate *you creep up to the PRF on tip-toes*. Do not get any closer than you have to in order to get a close-focus on the PRF entrance and never talk near the roost. Switch your mobile phone off as soon as you get out of the vehicle. If you are using an automated ultrasound detector that has audible output, switch the sound off or put it through headphones.

9.5.5 Remote-observation: Stage 5 – Observation

Remote-observation is very simple: keep the camera(s) in focus and operating continually. If you start the film with a shot of the time and date, after that the only pitfall is distraction, but you are not recording incidental bat activity going on in the background, so think like an Italian racing driver: *'what is behind you, doesn't matter…'*

Emergence

Emergence is the exiting of the PRF by a bat or a sequence of bats. Individuals do not give any warning of their emergence and do not typically stay in the vicinity of the PRF once they have emerged. Groups may signal that they are about to emerge with audible rasp-like squeaking (common to Daubenton's bats and noctules), or by scratching noises and hissing sounds (typical of noctules when grooming prior to exiting), or by an individual descending to listen at the roost entrance (typical of brown long-eared bats). However, overall, it is more common for there to be no warning of emergence in groups and, other than mating male noctules and Nathusius' pipistrelles, echolocation from a static position at the entrance is so rare as to be atypical.

Return

Although some groups may swarm outside a PRF immediately prior to sunrise, bats may return to the roost at any point throughout the hours of darkness. Individuals may forage in the early part of the night and return to the roost without emerging again that night. Mothers may emerge to forage, leaving non-volant young in the roost, but return to suckle their babies before emerging again to forage. These nursery returns are not in waves with groups of bats returning together, but comprise individual females returning at odd times throughout the hours of darkness in an erratic stream. Swarming aside, the typical return behaviour is similar to emergence; just as the group emerges in twos and threes, so they return in the same way, although the intervals between bats returning are greater than the intervals at emergence.

Visitation and Investigation

Visitation is the momentary entry of a PRF followed by immediate re-emergence. Visitation may take place at any point in the night, but does not result in the bat observed remaining in the PRF to roost over the following day.

Investigation is an approach by a bat to within 0.5 m of the PRF entrance, and typically right up to the entrance, but without fully entering. Investigation is typically made by an individual bat and may only comprise a single approach and immediate retreat, or a succession of approaches little more than a moment apart before the bat disperses. The bat may alight to one side of the entrance, but does not enter. Investigation is not a pass by a bat, even if that pass is within 0.5 m of the PRF; the bat must approach the entrance and then retreat.

Recording

Where bats are present you may try for an initial count, but it is more helpful to record the time the first bat emerged and the counter on the video camera screen so that the video may be more easily truthed back at the office.

It is not uncommon for a bat of another species that was roosting elsewhere to investigate the entrance of a PRF prior to a group within it emerging. Any such encounter should be recorded with the time so that ultrasound data can be checked to provide a species. This is anticipated on the recording form.

Where bats are not present, interpretation will fall to consideration of competitors and predators. The use of a zero-lux camera with a powerful infrared lamp allows the entrance of the PRF to be viewed with sufficient resolution to detect even invertebrate competitors such as woodlice, slugs and snails as they emerge to feed. Birds may also be observed returning to nest-holes and night-roost holes. Note that surveyors should be set up and situated sufficiently distant from the PRF as to not be obtrusive, and thereby discourage nesting or roosting birds from returning. Squirrels may also be observed, and although it has yet to be recorded, it is only a matter of time before someone records footage of a dormouse emerging. During a remote-observation survey, record absolutely everything that comes out of the PRF and anything that goes in.

9.6 Interpretation

Interpretation requires the consideration of the individual results in respect of each criteria (i.e. droppings, emergence, return, investigation, competitors/predators) and their cumulative meaning.

9.6.1 Presentation of observation results

The first stage of the interpretation should be to present the data in such a way as to make any anomaly that might have a positive bearing conspicuous. As a minimum, the objective questions at this stage would comprise the following:

1. *Were droppings discovered, and if so, how many?*
2. *Were bats encountered and was a conclusive account of their species, sex and numbers achieved?*
3. *If bats were encountered, is there any association between the species present and another rarer species? Refer to Table 8.4 in Chapter 8.*
4. *Was a bat observed investigating the PRF, and if so is the PRF of a type and in a situation the species is known to exploit?*
5. *Were competitors encountered that might support the hypothesis that it is 'more-likely-than-not' the PRF is suitable for exploitation by roosting bats, but conclusive evidence was not recorded?*
6. *Were predators encountered that might support the hypothesis that it is 'more-likely-than-not' the PRF is suitable, but is nonetheless not 'more-likely-than-not' to be exploited by roosting bats in this period?*
7. *Did the results of the individual inspections demonstrate changes, or were they constant?*
8. *Was no organism recorded emerging, investigating or returning to the PRF on any date?*[5]

The hypothesis achieved at objective question 4 would rationally be strengthened in likelihood if the answer to question 5 was also in the affirmative for the bat species (i.e. the species that investigated the PRF) and a competitor that was observed emerging, which are known to exploit the same internal environment (as illustrated in the species summary tables in Chapter 8).

Objective questions 1–3 are easily answered, but patterns in respect of questions 4–8 are less clear, and the interpretation that might satisfy them benefits from presentation. As the analysis is progressive, it might follow the format of Table 9.4, which makes any change from one visit to the next immediately obvious (a Word copy of this table can be downloaded from www.battreehabitatkey.com). This is of material benefit both during the surveillance and at the final interpretation of the results (see Stage 1 – Review).

9.6.2 Droppings

In most situations, any more than an individual dropping beneath the PRF would reasonably lead to the conclusion that a roost is nearby. It is vital that anything that looks like a dropping is collected for scrutiny with a hand lens, and if it is concluded that it is a bat dropping it is submitted for DNA analysis; the identity of the species can then be extrapolated to predict the period the roost is 'more-likely-than-not' to be occupied, the number of bats that might be present, and the function of the roost (overwintering, transitory, maternity or mating). Even the presence of an individual dropping will strengthen a hypothesis in combination with investigation and the presence of an associated competitor.

---

5   This would hint that the PRF is not a PRF at all, but simply a 'red herring'.

Table 9.4 A template for the presentation of repeat observation results for use at the interpretation stage

Tree reference No:					
DATE					
DROPPINGS					
EMERGENCE/ RETURN	Species				
	Numbers				
VISITATION	Species				
	Numbers				
INVESTIGATION	Species				
	Numbers				
COMPETITORS	Woodlice				
	Slugs				
	Snails				
	Wasp / hornet				
	Bees				
	Flies				
	Centipede / millipede				
	Bird				
	Squirrel				
	Mouse				

Constraints to interpretation in respect of droppings

As was identified in Chapter 8, a lack of droppings cannot be used to infer roost absence, but even where they are recorded, the discovery of droppings outside a roost requires careful interpretation and benefits from the application of a 'scepticality test'.

9.6.3 Emergence, return and visitation

Where emergence, return or visitation are encountered, any associations with another bat species should be identified by reference to Table 8.4 in Chapter 8 and the most up-to-date BTHK Database (see www.battreehabitatkey.com). Where associations are identified, these should be discussed and cross-referenced with any other data. For example, if an association between two or more species is identified, and these species have been identified as present during commuting or foraging surveillance, then there are grounds to suggest they may be 'more-likely-than-not' to exploit the PRF.

Constraints to interpretation in respect of emergence, return and visitation

Emergence, return and visitation can only be used as a positive. The fact that bats did not emerge, return or visit cannot be used in isolation to conclude unsuitability.

9.6.4 Investigation

Where investigation is recorded, the interpretation might identify whether the bat species observed is known to exploit that type of PRF in that sort of situation; and if so, when. If the result is positive for that season, but a competitor that is known to preclude roost presence was in occupation during the survey, then there are grounds to hypothesise further that the PRF is at least suitable as a roost. If the investigating bat species is found to exploit that PRF type but in a different season, then this will heighten the focus of any subsequent observation.

Where investigation is observed, associations with other bat species should be identified and cross-referenced accordingly.

Constraints to analysis in respect of investigation

Investigation behaviour can be tricky to interpret. It may be that the bat that investigated the PRF has roosted there on another occasion, or is prospecting the PRF for its suitability as a roost site (and may have found it unsuitable). It may be that the PRF is already occupied by the same or another bat species that did not emerge that night or return that morning. There may be a competitor or even a predator present. If no other organism is recorded, the observed investigation may have to rest as a 'loose end'.

9.6.5 Competitors

As was identified in Chapter 8, some competitors exploit the same environment as bats, and their presence may therefore suggest suitability of a PRF for an individual bat species; other competitors may preclude the presence of roosting bats year-round while they have possession of the PRF, and yet more may discourage bats from occupying a PRF in a given season.

Overall, the presence of competitors has a negative bearing upon the likelihood that bats will be present on the same day. This is illustrated at Table 9.5, where the odds in favour and against bats being present on the same day as a competitor species are identified. However, the fact that all the most frequently encountered competitor species have been recorded in tree-roosts on the same day that bats were present, in at least one period, demonstrates that there are associations and these associations might be reasonably used to infer the environment is suitable for roosting bats (although whether or not the internal dimensions are suitable may not be known).

9.6.6 Summing-up

The final analysis is as follows:

» If emergence, return, visitation (i.e. a bat fully entered the PRF) and/or droppings were recorded, proceed to Scenario 1.

» If investigation and an associated competitor recorded (i.e. a bat approached to within 0.5 m of the PRF entrance), proceed to Scenario 2.

» If neither emergence nor visitation was recorded, but a competitor species that is known to exploit the same environmental conditions as the target bat species was recorded, proceed to Scenario 3.

» If emergence, return, visitation, investigation or competitors were not recorded, and sufficient survey effort has been expended to test the hypothesis defined at the intelligence-gathering, desk-study and truthing stages (in line with the advice in Chapter 11), proceed to Scenario 4.

Scenario 1: Emergence, return, visitation and/or droppings

If emergence, return, visitation (i.e. a bat fully entered the PRF) and/or droppings were recorded, the PRF is a confirmed roost. If any more than an individual dropping was recorded below the PRF, then there is a *"reasonable likelihood"* the PRF is a roost.[6] A

---

6   Where only an individual dropping is recorded, the interpretation will require a more cautious approach and careful consideration of any additional evidence (visitation, associated competitor, etc.). Where a EPSL is being considered, it may be wise to seek advice from the Discretionary Advice Service offered by the appropriate government agency, prior to application.

Table 9.5 The % odds in favour (identified as + and in black) and against (identified as − and in red) associated with the most frequently encountered competitor species and the presence of roosting bats in each of the six seasons

Competitor	Season					
	Winter (Jan/Feb)	Transitory (spring-flux Mar/Apr)	Maternity		Mating (Sep/Oct)	Transitory (autumn-flux Nov/Dec)
			Pregnancy (May/Jun)	Nursery (Jul/Aug)		
Woodlice	+11% −89%	+12% −88%	+16% −84%	+7% −93%	+18% −82%	+11% −89%
Tree/dusky slugs	+5% −95%	+7% −93%	+6% −94%	+6% −94%	+10% −90%	+7% −93%
Garden snail	+8% −92%	+3% −97%	+2% −98%	0 −100%	+1% −99%	+2% −98%
Lipped snail	+1% −99%	+3% −97%	+1% −99%	+1% −99%	+3% −97%	+1% −99%
All spiders	+15% −85%	+13 −87%	+13% −87%	+13% −87%	+5% −95%	+16% −84%
Wasps and hornets	0 −100%	0 −100%	+0.6% −99.4%	+0.3% −99.7%	+0.2% −99.8%	+0.2% −99.8%
All bees	+0.3% −99.7%	0 −100%	0 −100%	0 −100%	0 −100%	+0.2% −99.8%
Millipede	+1% −99%	+2% −98%	0 −100%	0 −100%	+3% −97%	+1% −99%
Centipede	0 −100%	+1% −99%	+1% −99%	+1% −99%	+2% −98%	0 −100%
Blue tit	+3% −97%	+3% −97%	+3% −97%	+1% −99%	+1% −99%	+3% −97%
Grey squirrel	0 −100%	+1% −99%	0 −100%	0 −100%	+1% −99%	0 −100%
Common dormouse	0 −100%	0 −100%	+1% −99%	0 −100%	0 −100%	0 −100%

European Protected Species Licence (EPSL) will therefore be required if there is any potential that the roost might be destroyed or damaged, or bats disturbed, as a result of any proposed action. When considering the potential for damage and disturbance, it is sensible to consider not only direct impacts, but also indirect impacts (see the Indirect Damage and Disturbance, and Comparative Assessment of Environment discussion in Chapter 12).

Scenario 2: Investigation and an associated competitor recorded

If investigation and an associated competitor were recorded (i.e. a bat approached to within 0.5 m of the PRF entrance), then it may be concluded that there is a *"reasonable likelihood"* that the PRF is a roost. If any other surveillance concurrent with the roost surveillance (such as a passive ultrasound surveillance) has recorded another bat species that is known to exploit the same roost sites of the species encountered visiting, then it may reasonably be extrapolated that there is the potential that they will also exploit the same roost feature (if it is indeed suitable). However, without conclusive evidence, an application for an EPSL may not be appropriate. Nevertheless, it might be sensible to set out the evidence recorded with an appropriate impact assessment that assumes the PRF is a roost, alongside a due-diligence safeguarding strategy (such as that set out in the section dealing with Inconclusive and Null Results in Chapter 12), and seek advice from the Discretionary Advice Service offered by the appropriate government agency.

Scenario 3: Only an associated competitor recorded

If neither emergence nor visitation was recorded, but a competitor species that is known to exploit the same environmental conditions as the target bat species was recorded, then it may be concluded there is 'potential' that the PRF is suitable and might be a roost. Despite the perceived low level of risk, if there is any potential for damage to the PRF, it may be sensible to offer a generic due-diligence safeguarding strategy (such as that set out in the section dealing with Inconclusive and Null Results in Chapter 12).

Scenario 4: Null result

If emergence, return, visitation, investigation or competitors were not recorded, and sufficient survey effort has been expended to test the hypothesis defined at the intelligence-gathering, desk-study and truthing stages (in line with the advice in Chapter 11), then a reasoned argument will be required if the feature is to continue to be regarded as a PRF. Notwithstanding, in any situation where there is an interval between a null surveillance result and operations that will result in the loss or damage to a tree or larger area of wooded habitat, it is sensible to take a cautious approach and offer a due diligence safeguarding strategy. The reasons for this, along with a generic strategy, are provided in the section dealing with Inconclusive and Null Results in Chapter 12.

CHAPTER 10

Static-Netting

In this chapter

Introduction	The data the method is to collect
Equipment	The equipment needed to collect the data
Health and Safety	Considerations in respect of biotic hazards
Static-netting method	How to collect the data
Interpretation	How to present and interpret the data

10.1 Introduction

Static-netting comprises the capture of bats as they emerge from the roost, using a net that is held in a fixed position over the PRF entrance. Static-netting is typically used when bats are known to be present, but an aspect of their status is lacking, such as their species, their sex or the number of bats present.

> **Note:** In the UK the use of a static-net to catch bats requires a survey licence that can only be issued by:
> » Natural England.
> » Countryside Council for Wales.
> » Scottish Natural Heritage.
> » Northern Ireland Environment Agency.

> **Note:** Static-netting is not appropriate where there is any possibility heavily pregnant bats or when bats with dependent young might be present – typically June through mid-July (Collins 2016).
> **DO NOT TAKE RISKS.**

10.2 Equipment

For static-netting to be fully effective and the results reliable, it will in all but the most exceptional cases require:

» A ground-truthing account.
» A BTHK recording form.
» The static-net itself.

» A ladder (if PRF is safely accessible).

» Bat-handling gloves.

» Bat holding-bags.

» Identification texts and equipment.

» Portable shelter.

» A zero-lux video camera and infrared lamp or array of lamps sufficient to illuminate the full extent of the PRF entrance and at least 0.5 m on either side.

» An automated ultrasound 'bat-detector' that can be left unattended and will record in full-spectrum.

» A GPS.

» Some means of telling the time.

» Some means of telling what the temperature was at the start and end of the observation period.

» A head-torch that is sufficiently powerful to view the net frame in its entirety at the height it is to be deployed.

Apart from the recording form, static-net, bat-handling gloves, bat holding-bags, identification texts and equipment, and the portable shelter, all the equipment required and its use has been described in detail in the preceding chapter.

The BTHK static-netting recording form
The recording criteria adopted for the BTHK project were defined specifically to focus attention on the meaningful data required to fulfil the objective, see Table 10.1 (a Word copy of this table can be downloaded from www.battreehabitatkey.com).

When submitting a record to the Database all the attributes should be assigned a value.

The static-net
At present, static-nets come in three forms:

» Circular (the typical butterfly net).

» Triangular (also typically used for invertebrate surveying).

» Kite.

Despite an excellent design having been illustrated in the *Bat Workers' Manual* (Mitchell-Jones and McLeish 2004 – see Figure 4.1 on p. 42), up until very recently the only nets that were readily available were those that were designed for catching invertebrates. Unsurprisingly, butterfly-nets have historically been most favoured by bat surveyors but are only available 'off-the-shelf' with a typical maximum 60 cm diameter frame, and none have an internal clear PVC lip which is vital for static-netting from a PRF at height.

This need to adapt kit is also necessary where the target PRF has a longitudinal entrance, as the bats are well aware of the presence of the net and would simply exit above or below it. Competence at sewing together several net-bags was therefore required. In addition, in order to deploy the net over a PRF at height, a telescopic handle was also required. In practice, even where the PRF was accessible from the ground, without an extension to the handle, it was wise to use one and install the net with the handle fixed to avoid fatigue and any wobbling.

However, all this rather ad hoc cobbling together of kit is now unnecessary as Steaphan Hazell and his team at NHBS Ltd[1] have produced a static-net for the BTHK project that

1   NHBS Ltd, 1–6 The Stables, Ford Road, Totnes, Devon TQ9 5LE – www.nhbs.com.

Table 10.1 The BTHK static-netting recording form

SURVEYOR NAME			
DATE			
TREE LOCATION	OS Grid ref:		
CAMERA LOCATION	OS Grid ref:		
TIMING			
SURVEY COMMENCED			
SUNSET TIME			
SURVEY COMPLETE			
CONDITIONS UPON COMMENCING			
CLOUD COVER			
WIND			
RAIN			
TEMPERATURE			°C
CONDITIONS UPON COMPLETION			
CLOUD COVER			
WIND			
RAIN			
TEMPERATURE			°C
RESULT			
DROPPINGS PRESENT?			
EMERGENCE ENCOUNTERED?			
TIME FIRST BAT EMERGED			
SPECIES:			
COUNT:	♂	♀	

was purpose-built for tree-roosts. The net features a large kite frame, on a two-stage hinged bracket that allows the net to be deployed horizontally or vertically. The net has a clear PVC lip which together with the hinged kite frame facilitates a good seal to cylindrical and flat surfaces. The PVC also extends down into the bag in the same way as the curtain on a harp-trap, ensuring the bats do not escape before they are identified, and the net is white so the bats are clearly visible using a zero-lux camera. The handle can be extended by adding sections to 8 m, and is easily and safely deployed by two people. Once in place the handle end can be fixed to ensure there is no movement in the net position. Upon completion, the net can be lowered by two or more people, either drawing the handle away from the tree, or removing sections like a chimney-sweep's brush. Figure 10.1 shows the NHBS/BTHK static-net.

Ladder
In situations where static-netting is being performed to confirm the species or sex of bats already known to be present, if the PRF can be safely accessed, it may be possible to retrieve captured bats from the net bag without taking the net down by using a ladder.

Bat-handling gloves
Bats bite; it is good practice to wear protective gloves suitable for the size of the species and their level of aggression.

Figure 10.1 The NHBS/BTHK tree-roost-specific static-net. Top: the net is easy to deploy and the concave frame can be adjusted to suit the cylindrical substrate. The plastic membrane allows bats to crawl up behind but not escape the net, and the white net means that the bats can be seen using a zero-lux camera and infrared lighting configuration. The net can be easily removed from the frame for hand-washing and repair/replacement.

Bat holding-bags

A ready supply of sequentially numbered bat-holding bags will be required for the temporary storage of captured bats while they are individually identified, sexed, aged and released. These comprise a small cloth bag with a drawstring that can be tied-off round the neck of the bag, to prevent the bat escaping, and then hung-up (never placed on the ground) to ensure the bat is kept away from ground temperatures and damp, and safe from harm.

Identification texts and equipment

Regardless of the level of expertise, it is wise to carry an identification text in order to ensure that any uncertainty can be resolved. At the time of writing, the most up-to-date and complete key is provided in:

> » Dietz C and Kiefer A 2016. *Bats of Britain and Europe*. London: Bloomsbury.

However, the BTHK Project also still carries and makes regular use of the following (deferring to Dietz and Kiefer (2016) for the separation of *Myotis* and *Plecotus* spp.):

> » Stebbings R, Yalden D and Herman J 1986. *Which Bat Is It? A Guide to Bat Identification in Great Britain and Ireland*. London: The Mammal Society.

Calipers are required for measuring forearm lengths, and, in the case of long-eared bats, thumb lengths and tragus widths.[2] A digital camera capable of macro-photography is vital for getting close-up views of dentition, etc., as the image can immediately be enlarged and reviewed by several surveyors without increased stress to the bat. A supply of sample-tubes is also useful for situations where identification is uncertain; simply retrieve fresh droppings from a holding-bag and submit for DNA analysis. In all cases, a good-quality head-torch is required in order that all surveyors have both hands free to assist.

Portable shelter

Natural England (2013) recommend the use of a portable shelter in which bats can be processed in as warm environment as possible. A 'pop-up festival tent' is ideal, and provides the required warm environment and ensures that all equipment is together in one location and away from the attentions of invertebrates.

10.3 Health and Safety

All surveyors must have up-to-date inoculation against European Bat Lyssaviruses (EBLVs) 1 and 2, commonly referred to as bat rabies (see also discussion in respect of dropping samples at Chapter 8): **DO NOT TAKE RISKS**.

In addition, it is up to the individual team to ensure they have insurance and have performed a risk assessment. However, it cannot be overstressed that insect-repellent should be liberally applied (taking care not to get it on bat-handling gloves or holding-bags) and the survey team should ensure that they have sufficient rest before and after a netting survey for them to be safe to drive and within legal limits with respect to work-breaks, in accordance with the most up-to-date legislation.

10.4 Static-netting method

As with a remote-observation survey, static-netting is not simply about catching bats, because the bats may not be present on the evening the survey is performed. Static-netting is about searching for all the clues that might hint that the PRF is or is not exploited as a roost.

A thorough survey will comprise six stages, as follows:

» **Stage 1** – Review.

» **Stage 2** – Assessment of conditions.

» **Stage 3** – Search for droppings.

» **Stage 4** – Set-up.

» **Stage 5** – Static-netting.

» **Stage 6** – Retrieval and handling.

By completing all six stages carefully, sufficient data will be provided for a meaningful interpretation.

Welfare

In order to ensure the potential for injury to any bat is kept at a minimum, the net should not be moved once it is deployed unless it is absolutely necessary to do so, and then the net must not be dragged across the tree, but lifted away and gently replaced; do not attempt this with only one person when the net is above 4 m. Once deployed, the net

---

2   Although grey long-eared bats are not known to roost in trees, the identity of any bat should not simply be assumed.

must be monitored continually using the camera and infrared lamp to view the tree and the net and ensure the position does not pose a hazard to any bat emerging.

Note: Static-netting should only take place at temperatures above 8°C unless specifically authorised under a project licence. Therefore, if the temperature drops to 7°C the net should be retrieved and the survey continue as remote-observation, and if the temperature drops still further the survey should be abandoned.

Constraints
A full list of constraints is provided in Chapter 7, but in summary the primary constraints associated with the efficacy of static-netting are that:

» The method cannot detect droppings or changes to the internal substrate.

» Not all bats observed can be identified to species.

» Even where a group is observed, the sex of some species cannot be inferred.

10.4.1 Static-netting: Stage 1 – Review
The first stage of every observation should be the review of the results of the ground-truthing, and any pre-existing account(s) of earlier observation(s) or netting.

10.4.2 Static-netting: Stage 2 – Assessing the conditions
Weather conditions have been discussed in the preceding chapter. However, the temperature threshold is higher for static-netting, 8°C, and increased diligence is required in terms of:

» Ensuring the net has fixed-stability once deployed, and the bag is truly 'static' and not flapping about due to wind.

» Ensuring that the net does not become even so much as slightly damp.

If there is wind, the tree will move and the net will not be stable. In most situations, it will be impossible to deploy the net in anything other than calm conditions, but if the survey team are not diligent, and do not maintain focus on the net and miss movement as the temperature drops at sundown (and there is always some movement), there is a risk of bats being crushed between the frame and the tree or the net falling, with catastrophic results to any bat caught. Regardless, if there is sufficient wind to move the tree, there will be sufficient movement to flap the net and any bats present will be deterred from emerging. Welfare principles also apply in respect of rain or any other atmospheric moisture. It is up to the survey team to judge the conditions, but absolute integrity is required; if there is any doubt, take the net down and either revert to remote-observation alone or abandon the survey.

10.4.3 Static-netting: Stage 3 – Search for droppings
Even though the PRF may not be accessible, before the net is deployed, the first stage of any static-netting survey should still be to perform a thorough search for droppings on the ground beneath the PRF, and on any field and accessible shrub-layer vegetation, etc. (see Chapter 8, subsection 8.5.3).

10.4.4 Static-netting: Stage 4 – Set-up
Stage 4 should proceed as follows:

1. Arrive at the site sufficiently early that having walked to the target tree, the surveyors will have no less than 1.5 hours to set-up and then acclimatise to the situation and take in their surroundings prior to the time the first bat might emerge.

2. Regardless of the height the net is to be deployed at, clear the ground of any obstructions to ensure the net can be deployed and retrieved in one smooth movement.

3. Raise the net to the level of the entrance but do not deploy it before the position the handle will be secured has been agreed, and retrieval has been tested, so that everyone is satisfied as to what their role is in the operation.

4. Deploy the net by raising it, placing the handle securely on the ground in the position chosen, and using this as a fulcrum to slowly and gently place the net over the PRF entrance with the net hanging entirely free of any part of the frame, so that any bat falling can only come into contact with the net.

5. Retreat.

6. Take into account that until the light levels drop sufficiently, it may not be possible to switch the camcorder to night-shot mode or focus the infrared lamp accurately upon the PRF – adjustments will have to be made as the observation progresses.

7. Take into account the fact that the PRF may begin to move, and, where possible, choose an observation position where the camera and lamp can be adjusted to a wider field of vision without redeployment in another location, in case the static-netting aspect of the survey has to be abandoned and the survey resort to remote-observation alone.

8. Take into account the advice given in the previous chapter regarding noise.

10.4.5 Static-netting: Stage 5 – Netting
Follow the advice given in respect of the camera in Chapter 8.

It is helpful to divide responsibility as follows

» **Scribe** – responsible for:
1. Ground end of net handle.
2. Monitoring temperature (if it drops below 8°C retrieve the net and continue with remote-observation alone; if the temperature then falls below 7°C, the survey should be abandoned).
3. Reading identification keys.
4. Handing holding-bags to the 'handler'.
5. Completing the recording form.
6. Ensuring all bats are processed and released and all holding-bags are accounted for prior to packing up.

» **Handler** – responsible for:
1. Net end of net.
2. Retrieval of net.
3. Monitoring camera to ensure captures are acted upon promptly.
4. Removal of bats from net.
5. Handling bats for identification.
6. Taking all measurements.
7. Release of bats.
8. Redeployment of net (if required).

Natural England (2013) highlight the need to consider how many bats are likely to be caught and over what time period. This is because the length of the productive foraging period may be temporally limited, and the bats may therefore miss the peak due to being held in the net. Obviously, there is a balance between getting the required data, moving the net too often and negative effects upon the bats caught. It is up to the survey team to judge the conditions and the length of time between capture and release on a species-specific basis, but keep in mind that the longer a bat is in the net, the greater the probability that it will escape. As a general rule, if five minutes pass following a capture, it is sensible to take the net down, bag the bat and redeploy the net (if it is safe to do so).

In practice, one to five bats will provide sufficient age, sex and breeding condition information to give confidence enough to extrapolate as to the nature of the roost. In a typical situation that is achieved within a short period after the first bat has emerged. The net can then be retrieved and the count achieved using the zero-lux camera.

10.4.6 Static-netting: Stage 6 – Retrieval, handling, identification and release

Retrieval

Regardless of the deployment height, retrieval of the net requires a minimum of two people.

This is due to the need to lower the handle in a slow and controlled manner. To this end, one person will gradually manoeuvre the end furthest from the tree, as the other gradually works towards the net and takes charge of it. This process will bring the entirety of the handle under complete control in a broadly horizontal orientation. At this point the person furthest from the tree can also proceed to the net to assist with the release of trapped bats, and subsequent handling and identification.

At this point lighting is essential. If the net can be retrieved and redeployed confidently with a red filter on the headlight, then so much the better. However, the bats now know the net is there, so if white lighting is required, be sure the net is under control and no bat is harmed then this is the only option.

In situations where static-netting is being performed to confirm the species or sex of bats already known to be present and a ladder is deployed, one surveyor should maintain control of the handle from the ground to ensure control is maintained.

Handling

It is expected that anyone attempting to handle bats will be licensed by the appropriate government agency, and therefore sufficiently experienced to have attained confidence in handling bats, or will be closely supervised by someone else who is. However, for novices reading this book as part of their training, excellent advice is provided in:

» Mitchell-Jones A and McLeish A (eds) 2004. *The Bat Worker's Manual*, 3rd edition. Peterborough: Joint Nature Conservation Committee.

In accordance with good practice, all bats should be removed from the net and placed individually in their own holding-bag, and the bags hung in a safe area, ideally within a portable shelter (see Kunz *et al.* 2009; Natural England 2013; Collins 2016). All bats should thereafter be identified immediately so there is the least impact in terms of distress and foraging time loss.

Note: It is vital to number the bags and use them in that order. It is recommended that one person acts as 'scribe' and the other acts as 'handler' and these roles are not swapped at any point during the trapping period. The scribe will therefore keep an accurate record of how many bags are 'occupied' to ensure all bats are released when processing is complete.

Identification and recording

The following information is required:

» Species.
» Sex.

» Reproductive condition.

» Age.

» Overall number.

This also accords with the information identified by Kunz *et al.* (2009), who provide useful advice regarding how to collect this data.

Note: If a pregnant bat is captured, at the moment this is discovered netting should cease and all bats netted should be released without identification or any other processing.

Release
In most situations, release can be performed by orientating the bat in the hand to allow it to be belly down on the surveyor's fingers with the head facing out above the surveyor's forefinger. The arm is then extended with the hand at head-height, thumb uppermost, little finger lowest. The grip is then relaxed allowing the bat to climb up onto the fore-finger. If the bat is sufficiently warm to take wing, it will begin to release itself and in most cases, will simply fly from the forefinger, but if it is reluctant to do so, place the thumb across the bats shoulders and gradually open the hand to offer a horizontal surface, and then move the thumb away and hold the open hand flat, but raise the arm as high as possible.

Do not ever attempt to assist the bat in taking wing. If it is reluctant to take wing from a surveyor's hand, the bat should be released onto the roost-tree or an adjacent tree where there is an open flight-path, and then the survey team should retreat and leave the release position in complete darkness.

10.5 Interpretation
The interpretation is identical to that for remote-observation, as set out at Chapter 9.

CHAPTER 11

Surveillance Effort

In this chapter	
Introduction	Acknowledgement of variables and data shortages
Detectability variables	Distribution and abundance, seasonal usage of trees, activity, sex and duration of occupancy
Surveillance periods and intensity therein	Winter, spring-flux, pregnancy, nursery, mating and autumn-flux

11.1 Introduction

The variables that decide how much surveillance effort might be proportionate to support a robust impact assessment are manifold, and at the time of writing the detectability of the individual bat species in certain periods of the year is imperfectly understood.[1]

Nevertheless, British and European studies have provided sufficient data for a threshold of *"reasonable likelihood"* in respect of the probability of encounter to be calculated for Bechstein's bat in the pregnancy and nursery periods, and the evidence collated supports the suggestion that the same threshold will be broadly similar for the barbastelle, Daubenton's bat, Natterer's bat and brown long-eared bat.

Furthermore, although reliant upon a small sample of 'low-resolution' monthly recordings, an attempt has been made to provide a degree of evidence to direct surveillance effort in respect of individual bats in the pregnancy and nursery periods, and all numbers in the winter, spring-flux, mating and autumn-flux periods.

Notwithstanding, it should be noted, that all bat species and all habitats have had to be 'lumped' at this stage, and a threshold of *"reasonable likelihood"* still cannot be defined in respect of surveillance intensity for: Brandt's bat; the whiskered bat; maternity groups of *Nyctalus* spp. and *Pipistrellus* spp.; and all male groups in the pregnancy and nursery periods.

It is accepted that this is a significant failing that must be acknowledged. Even then, this chapter must open with a warning:

> While all the thresholds identified in this text are based on published accounts in peer-reviewed white papers and records held on the BTHK Database, they have not been subject to any controlled test.

---

1 There are people who are of the opinion that some questions cannot be answered. The BTHK project's opinion is that 'grey areas' simply demonstrate a failure to commit to the objective.

All the suggestions are rational, but it must nonetheless be stressed that they do not represent a prescribed set of rules. Therefore, if the surveillance team have evidence that an alternative course of action might yield more meaningful data, they should have faith in their own judgement and proceed accordingly.

11.2 Detectability variables

11.2.1 Overview
A summary of detectability variables that might be worth consideration comprises:

1. The distribution of the target (i.e. some bat species do not occur in all regions of the UK).
2. The abundance of the target (i.e. there are more bats of some species than there are of others).
3. The seasonal usage of trees by the target (i.e. not all bat species occupy trees in all seasons for the same purposes so some will be absent in certain periods).
4. The sex of the target (behavioural differences that separate males and females).
5. Activity (i.e. whether the bat is active and therefore potentially mobile during the hours of darkness, or torpid and therefore more likely to be sedentary for a period greater than 24 hours).
6. The duration the target stays in one location (i.e. how long individual species and the different sexes occupy a PRF before moving to another).
7. The methods available to locate the target.

11.2.2 Distribution and abundance
Only five bat species present in the UK occur in all regions: Daubenton's bat, Natterer's bat, common pipistrelle, soprano pipistrelle and the brown long-eared bat. It might therefore be rational to assume the odds of encountering a Natterer's bat in a tree should be broadly equal to those of encountering a brown long-eared bat. However, although they have broadly the same distribution, their abundance (amongst other factors) is immediately identifiable as a significant difference between the two species if we accept the population estimates that suggest there are 60% more brown long-eared bats than there are Natterer's bats.[2]

11.2.3 Seasonal usage of trees
Referring back to Table 2.3, 'The bat year', in Chapter 2, it is immediately evident that not all bat species occupy trees year-round. The situation is simplified in Table 11.1, which shows the maximum number of species that might potentially be present in each of the six periods and their individual months.[3]

At a glance, it looks like the highest probability of encountering a bat is in April and June and the lowest is in January and February, but this is misleading.

In fact, pipistrelle females are generally absent from the trees, and the females of the species that are present are grouped in 'maternity colonies', and are in any case numerically less abundant than the pipistrelles. What this means is, that although more *species*

---

2 In practical terms, long-term surveillance of populations of both species have demonstrated that in Wytham Woods, Oxford there are three brown long-eared bat social-groups to every one of Natterer's bats, which suggests the estimates are broadly accurate (D. Linton personal communication April 2016).
3 Note that five of the species have significantly restricted distributions, which means that in much of the British Isles the numbers will be lower.

Table 11.1 The maximum number of bat species that might potentially be present in each of the six periods and their individual months

The bat year	WINTER		SPRING-FLUX		PREGNANCY		NURSERY		MATING		AUTUMN-FLUX	
	Jan	Feb	Mar	Apr	May	Jun	Jul	Aug	Sep	Oct	Nov	Dec
Number of bat *species* occupying trees				13	12	13			12	12	12	
							11	11				
			9									9
	8	8										

may be present, as the populations of these species are significantly smaller, there may therefore be fewer bats present overall, and those that are present are in any case occupying fewer trees because they are grouping together. Therefore, the probability of encountering them is lower.

11.2.4 Activity

In the context of discovering a tree-roost, activity relates to whether the bats are following an even 24-hour cycle of sleeping and waking, or are following a cycle with longer periods of torpor interspersed with irregular periods of waking. While it is not particularly scientific, it is rational to suppose that the more torpid the bat is, the less likely it might be that it will move, and therefore the longer each duration of occupancy might be. Figure 11.1 illustrates the proportion of encounters with bats of all species that were awake or torpid in each month in records held on the BTHK Database.

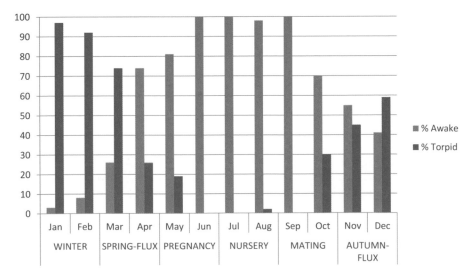

Figure 11.1 The proportion of bats that were awake and torpid in each period in records held on the BTHK Database.

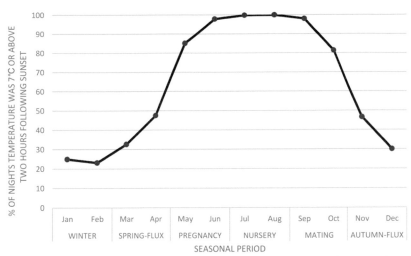

Figure 11.2 The % of nights in each period that the temperature was 7°C or above two hours following sunset. Data provided by the Met Office.

It is evident that the months of greatest torpor are January and February when only noctules and pipistrelles have been encountered awake (and the former have been notably drowsy), and the months of least torpor are June–September, when only barbastelles have been encountered in apparent torpor.

However, when considering activity, it is also important to look at the preceding period. For example, grouping the autumn-flux data to the low resolution of both months and all species, and doing the same with the winter data, demonstrates that 48% of bats encountered during the autumn-flux period were awake but this had dropped to only 6% during the winter period.

There is a perception of an en masse roost-switch between the autumn-flux and winter periods. In fact, the BTHK Database holds accounts of an overall 179 paired inspections for December and the following January. Of these, bats were encountered in 29 situations, comprising:

» 15 accounts of occupancy in December but vacancy in the following January inspection.

» One account of vacancy in December but occupancy in the following January.

» 13 accounts of occupancy by the same species in both inspections.

Although the data-set is small, it does demonstrate that 55% of bats switched roosts between December and January, and of this proportion, 92% moved to a roost that was outside the surveillance zone. Looking at the Met Office analysis used in Chapter 9 from a different angle, the percentage of nights that the temperature two hours following sunset was above 7°C is still significant in the winter period (as demonstrated by Figure 11.2). Perhaps this switch is to cooler quarters in order to remain torpid when winged prey species are overwintering as larvae, and otherwise unavailable.

11.2.5 Sex

The way the sexes behave towards each other has a bearing on how detectable they may be due to whether they roost together or apart, and whether the males advertise or not. In simplistic terms, the year can be split into three periods:

1. **Attraction** – During the mating season and into the autumn-flux the sexes are attracted toward one another. However, although the males of the *Nyctalus* spp. and *Pipistrellus* spp. may make the presence of a roost more conspicuous with their advertisement calling, the barbastelle, *Myotis* spp. and brown long-eared bat do not.

2. **Tolerance** – During the winter and into the spring-flux both sexes of the same species may be tolerant of each other in communal roosts.

3. **Repulsion** – During the pregnancy and nursery periods the females repulse the males of the same species and adult males do not typically share roosts with maternity groups.

11.2.6 Duration of occupancy

In the most basic terms, during the colder periods – spring-flux, autumn-flux and winter – all bat species are at their most sedentary, and during the warmer periods – pregnancy, nursery and mating – all bat species are at their most mobile.

However, mobility between roosts (i.e. roost-switching) and therefore the duration of occupancy of an individual PRF is also stratified by sex, with females more mobile than males. The females also exhibit a gradation in mobility, with maternity groups most mobile between roosts during early pregnancy, and gradually less mobile when young are born and cannot fly and must therefore be carried.

Table 11.2 sets out a summary of a review of the durations of roost occupancy exhibited by female groups at this time.

There is one important question that may be asked of all the data summarised in Table 11.2: does it relate to an individual bat, or does it relate to the entirety of the group. To elaborate: was the occupied PRF unoccupied the day before the tagged bat led the study team to it? Was the PRF entirely abandoned the day the tagged bat switched? Or, was the move more gradual, with some bats arriving before the tagged bat joined them, the numbers then fluctuating from one day to another, and then the numbers dwindling as the overall colony moved on? Only long-term surveillance of the PRF itself rather than the bats, would answer the question of duration of occupancy.

Records held on the BTHK Database are presented to cover the periods for which no occupancy data is reported in white papers. The same is true for individuals in the pregnancy and nursery period. This data has a very important caveat: all of it was gathered using the close-inspection method and we really do not know how disturbing the intrusion by an endoscope actually is. If there is a significant displacement effect, is there a gradation between sexes and species?

11.3 Surveillance periods and intensity therein

11.3.1 General

If a *"reasonable likelihood"* of roost presence has been established at the ground-truthing stage, this will have established within which of the six search periods surveillance would be proportionate (i.e. one or more of: winter, spring-flux, pregnancy, nursery, mating and autumn-flux).

Height data will also logically influence when the PRF will be surveyed. For example, looking at the height data presented in Chapter 2, on the basis of records held on the BTHK Database at the time of writing, there are no grounds to predict roost presence in

Table 11.2 A review of published scientific and anecdotal evidence in respect of roost-switching behaviour exhibited by tree-roosting bat species

Barbastelle *Barbastella barbastellus*

Roost-switching interval	On average, every 4 days (range 3–5) (Billington 2004) Frequently moves roosts after only a single day's occupancy (Greenaway 2008) On average, every 1.6 days (Kerth and Melber 2009) On average occupancies lasting 2.3 days within a range of 1–5 (Zeale 2011) Males moved on average every 4 days (Russo *et al.* 2005)
Number of trees colony distributed between	3–5 (Greenaway 2008)
Total resource of trees needed in a single year	27 trees (Kerth and Melber 2009)

Bechstein's bat *Myotis bechsteinii*

	Pregnancy	Lactation	Weaning
Roost-switching interval	1.5 days (Dietz and Pir 2011)	3.3 days (Dietz and Pir 2011)	4.8 days (Dietz and Pir 2011)
	Males switch an average of every 4.1 days in May but became more sedentary as the summer progressed (Dietz and Pir 2011)		

	Pregnancy	Lactation	Weaning
Number of trees colony distributed between	2 (Dietz and Pir 2011)	2 (Dietz and Pir 2011)	3 or more (Dietz and Pir 2011)

Total resource of trees needed in a single year	35–40 (Dietz and Pir 2011)

Alcathoe's bat *Myotis alcathoe*

Roost-switching interval	No data
Number of trees colony distributed between	No data
Total resource of trees needed in a single year	No data

Brandt's bat *Myotis brandtii*

Roost-switching interval	4–11 days (Sachanowicz and Ruczyński 2001) Frequently (Dense and Rahmel 2002)
Number of trees colony distributed between	No data
Total resource of trees needed in a single year	No data

Daubenton's bat *Myotis daubentonii*

	Pregnancy	Lactation	Weaning
Roost-switching interval	2 days (Lučan and Radil 2010)	3 days (Lučan and Radil 2010)	4 days (Lučan and Radil 2010)
	Switch on average every 2–5 days (Encarnação *et al.* 2005) Every 2 days within a range of 1.1–3.5 (August *et al.* 2014)		

Number of trees colony distributed between	No data
Total resource of trees needed in a single year	Up to 40 trees in a single year (Geiger and Rudolph 2004)

Table 11.2 – continued

Natterer's bat *Myotis nattereri*	
Roost-switching interval	Switch on average every 3 days (Smith and Racey 2008) Every 1–4 days (Simon *et al.* 2004) Every 2–5 days (Dietz *et al.* 2011)
Number of trees colony distributed between	One during pregnancy and lactation, more thereafter (Smith 2001)
Total resource of trees needed in a single year	No data

Whiskered bat *Myotis mystacinus*	
Roost-switching interval	Potentially every 10–14 days (Dietz *et al.* 2011)
Number of trees colony distributed between	No data
Total resource of trees needed in a single year	No data

Leisler's bat *Nyctalus Leisleri*	
Roost-switching interval	Potentially every 3 days (Fuhrmann *et al.* 2002)
Number of trees colony distributed between	No data
Total resource of trees needed in a single year	Up to 50 trees in a single year (Schorcht 1998; Meschede and Heller 2000)

Noctule *Nyctalus noctula*	
Roost-switching interval	Potentially every 10–15 days (Pénicaud 2000)
Number of trees colony distributed between	No data
Total resource of trees needed in a single year	Up to 60 trees used in a single year (Frank 1997)

Nathusius' pipistrelle *Pipistrellus nathusii*	
Roost-switching interval	No data
Number of trees colony distributed between	No data
Total resource of trees needed in a single year	No data

Common pipistrelle *Pipistrellus pipistrellus*	
Roost-switching interval	Potentially every 11–12 days (Feyerabend and Simon 2000)
Number of trees colony distributed between	No data
Total resource of trees needed in a single year	No data

Soprano pipistrelle *Pipistrellus pygmaeus*	
Roost-switching interval	No data
Number of trees colony distributed between	No data
Total resource of trees needed in a single year	No data

Table 11.2 – continued

Brown long-eared bat *Plecotus auritus*	
Roost-switching interval	Potentially every 2–4 days (Heise and Schmidt 1988; Fuhrmann 1991; Fuhrmann and Seitz 1992)
Number of trees colony distributed between	No data
Total resource of trees needed in a single year	No data
Lesser horseshoe-bat *Rhinolophus hipposideros*	
Roost-switching interval	No data
Number of trees colony distributed between	No data
Total resource of trees needed in a single year	No data

the ground-layer (i.e. below 0.5 m) in the pregnancy, nursery or mating periods. What that means is (at least in a commercial situation), whilst the surveillance team might look in such features if they were contracted to look at others higher up in the canopy at another time, if all that was present were PRFs in the ground-layer, some justification would be necessary if it was recommended that ground-level PRFs were to be visited outside the periods in which there was evidence that they were exploited.[4]

The Natural England Licensing Consultation suggests that in situations where *'the ecological impacts of the development can be predicted with sufficient certainty and mitigation or compensation will ensure that the licensed activity does not detrimentally affect the conservation status of the local population of any European Protected Species, the costs or delays associated with carrying out standard survey requirements would be disproportionate to the additional certainty that it would bring'* (Natural England 2016).

Therefore, before assuming surveillance will be justified, it should be established whether the impacts cannot be defined and understood in the context of the bat species potentially present without surveillance data, and whether there are proven, effective methods to mitigate the impacts, or compensate for them with habitat provision elsewhere and the provision of artificial roost-boxes (i.e. does the bat species adopt bat-boxes for the purpose that the PRF might be exploited). If the answer to both questions is yes, then the rationale for performing the surveillance in each of the six periods should be questioned. Furthermore, where surveillance is justified the following caveat applies:

> Although a threshold of *"reasonable likelihood"* has been attempted for all periods, the variables that will influence the surveillance intensity confound the production of comprehensive written advice. As a result, in many situations the intensity will have to be defined by the survey team using their own judgement following reference to the situation and species ecology. Nevertheless, an effort has been made to offer some support at the close of each period account under the heading of 'Proportionality'.

---

4   The most up-to-date database should be reviewed to ensure this advice is still correct: see www.battreehabitatkey.com.

11.3.2 Winter

Number of bat species potentially present

Eight bat species have been recorded occupying trees in the winter period:

- » The barbastelle.
- » Natterer's bat.
- » The noctule.
- » Leisler's bat.
- » Nathusius' pipistrelle.
- » The common pipistrelle.
- » The soprano pipistrelle.
- » The brown long-eared bat.

Male and female behaviour

The bats that have been recorded in trees in winter period have not been sexed; it is therefore unknown whether there is any bias generally, or within any species. However, the general trend (at least in subterranean winter quarters) appears to be toward tolerance between the sexes at this time.

Activity

Unsurprisingly, winter is the time of lowest activity. Looking at the data presented in Figure 11.1 more closely, there is a gradation in how awake different species of bats are when they are encountered, as follows:

1. Natterer's bats and brown long-eared bats – Entirely torpid.
2. Noctules – Either entirely torpid or aware of the endoscope but notably drowsy in their response to the endoscope.
3. Common and soprano pipistrelle – 75% of encounters were with bats that were awake, aware and often aggressive/defensive in their response.

Duration of occupancy

Looking at repeat records held on the BTHK Database, there are an overall 183 paired records where the same PRF was inspected in January and February of the same year, comprising the following situations:

- » PRF unoccupied in both months – 168 instances.
- » PRF occupied in both months – nine instances, relating to Natterer's bat (one account), noctule (one account), common pipistrelle (two accounts), soprano pipistrelle (one account), and brown long-eared bats (four accounts).
- » PRF occupied in January alone – three instances, relating to noctule (one account), common pipistrelle (one account) and soprano pipistrelle (one account).
- » PRF occupied in February alone – three instances, relating to Natterer's bat (one account) and brown long-eared bats (two accounts).

"Reasonable likelihood"

Taking the BTHK Database evidence into account, 60% of roosts that were occupied in winter were occupied in both January and February by the same species. Although not particularly scientific, as over 50% of repeat inspections recorded the same situation, it might reasonably be argued that a single inspection[5] in the period late January through early February would be 'more-likely-than-not' to reflect the true situation, and therefore have a *"reasonable likelihood"* of detecting a winter roost.

Proportionality
Where surveillance is justified, the data available suggest a proportionate approach might be to do an individual visit in late January through early February.

Notes
Taking into account torpor and nightly temperatures, whether a meaningful survey can be performed in the winter period will depend upon whether a close-inspection is possible. This will itself depend upon: whether the PRF is safely accessible (both in terms of the surveyor, but also taking into account the fragility of many winter roost sites, and the vulnerability of the bats therein); and whether it can be comprehensively searched. In some situations, a survey may be impossible or impractical. For example, and where close-inspection was impossible, nothing more can be done. However, in any situation where access was deemed impractical it would be wise to guard against a charge of laziness, by identifying what action might have been taken to facilitate a close-inspection, and robustly argue why it was either irresponsible (such as attempting to investigate a loose bark-plate) or sufficiently impractical to be disproportionate to the level of risk.

> Be aware: blue tits are very much in evidence night-roosting in trees in winter, and their duration of occupancy appears to be continual throughout the winter period. In many cases, their presence is identifiable by a white dropping stain and a strong smell of peanuts in the PRF entrance.

Spring-flux

Number of bat species potentially present
Thirteen bat species have been recorded occupying trees by the close of the spring-flux period:

» The barbastelle.

» Bechstein's bat.

» Alcathoe's bat.

» Brandt's bat.

» Daubenton's bat.

» Natterer's bat.

» The whiskered bat.

» The noctule.

» Leisler's bat.

» Nathusius' pipistrelle.

» The common pipistrelle.

» The soprano pipistrelle.

» The brown long-eared bat.

---

5   Inspection may be taken literally, as close-inspection is the only surveillance method that will be reliable in the winter period.

Male and female behaviour

In spring, those species that were overwintering underground begin, by stages, to return to the trees. It has been suggested that males emerge before females, but that has yet to be proved. Regardless, by the end of the spring-flux, maternity groups of females have begun to form.

Activity

Accounts within the BTHK Database demonstrate that there is a notable change in the consciousness of bats between March and April. When all the records are aggregated, 76% of bats encountered in March were judged to be torpid, but in April the situation had reversed, with the same proportion now judged to be awake and aware of the endoscope.

Duration of occupancy

Looking at repeat records held on the BTHK Database, there are an overall 154 paired records where the same PRF was inspected in March and April of the same year, comprising the following situations:

» PRF unoccupied in both months – 115 instances.

» PRF occupied in both months – 12 instances, relating to Daubenton's bat (one account), Natterer's bat (one account), noctule (one account), common pipistrelle (two accounts), soprano pipistrelle (one account), brown long-eared bats (four accounts), and two changes of species (one account of a common pipistrelle being replaced by a Natterer's bat, and one account of a Natterer's bat being replaced by a brown long-eared bat).

» PRF occupied in March alone – 14 instances, relating to Natterer's bat (four accounts), noctule (one account), common pipistrelle (one account) and brown long-eared bats (eight accounts).

» PRF occupied in April alone – 13 instances, relating to Bechstein's bat (one account), Daubenton's bat (one account), Natterer's bat (four accounts), noctule (one account), common pipistrelle (two accounts) and brown long-eared bats (four accounts).

"Reasonable likelihood"

Taking the BTHK Database evidence into account, 74% of roosts that were occupied in the spring-flux period were occupied in one month alone (39% in March and 35% in April), leaving only 26% occupied in both months (and not always by the same species). A perfectly reasonable argument might be that at least two visits would be required to have a *"reasonable likelihood"* of detecting a spring-flux roost. However, in practice the number of visits in the spring-flux period may require the application of common sense.

If the instances an individual PRF was occupied in both months are added to the individual months, 52% of encounters were recorded in March and 48% were recorded in April, which suggests *"reasonable likelihood"* might be achieved by an individual visit in March, but the span of meaningful data is small and comprises two very different years in terms of the prevailing spring weather. In practical terms, it helps to keep in mind that the spring-flux is not simply a continuation of winter, but an identifiable end to it. Therefore, in some years it is sensible to delay any action until spring is clearly underway and this often does not manifest until April. Viewing weather data online,[6] in very general terms there is a 2°C temperature jump between February and March, and another 2°C jump between March and April, and although rainfall is broadly constant between February and March, it drops significantly between March and April. As a result,

---

6   Simply type *'annual weather trends by month in … (insert county of choice)'* into Google.

in some years it may be sensible to delay action until April, and make an individual inspection when the conditions can not only be predicted to be improved generally, but are also more stable.

Proportionality
Where surveillance is justified, the data available suggest a proportionate approach might be to do an individual survey in mid-March. However, in years where the weather conditions remain poor and spring is demonstrably late, the proportionate approach might be to delay until the weather is improved and perform the survey in April.

Notes
Where close-inspection is impossible/impractical, and the surveillance method will therefore comprise remote-observation, it should be noted that, working in Oxford, Danielle Linton has noted that in April there may only be a 2-hour window of suitable conditions in any one night, and bats may therefore emerge long after the typical 2 hours following sunset (D. Linton personal communication April 2016). As a result, it may be sensible to continue the survey for the entirety of the night.

11.3.3 Pregnancy

Number of bat species potentially present
Thirteen bat species have been recorded occupying trees in the pregnancy period:

- » The barbastelle.
- » Bechstein's bat.
- » Alcathoe's bat.
- » Brandt's bat.
- » Daubenton's bat.
- » Natterer's bat.
- » The whiskered bat.
- » The noctule.
- » Leisler's bat.
- » The common pipistrelle.
- » The soprano pipistrelle.
- » The brown long-eared bat.
- » Lesser horseshoe-bat.

Male and female behaviour
During the pregnancy period, the tolerance the sexes had for each other is temporarily at an end, and the males roost apart from the females. As was identified in Chapter 2, different species appear to adopt different strategies, with males of some species still thought to more frequently roost alone, but Daubenton's bats, Natterer's bats, noctule and Leisler's bats have also been encountered occasionally in all-male groups. The key point here is that the numbers of bats are more or less evenly divided between the sexes, so if there are 60 females occupying two or three trees, as with Bechstein's bat maternity groups, there may be individual males in 60 other trees and therefore anything up to 63 trees occupied. But if the male Daubenton's bats have decided to group, there may only be two trees occupied: all the females in one and all the males in another.

Activity

Reference to Figure 11.1 demonstrates that by the pregnancy period it is rare to encounter a torpid bat.

Duration of occupancy

The current knowledge in respect of the duration of occupancy for maternity groups in the pregnancy period is broadly defined for all species in Table 11.2.

However, to provide some degree of sensible advice in respect of individuals, paired records (i.e. where the same PRF was inspected in both May and June) were identified and the results collated. In total, there are 82 paired records on the BTHK Database, comprising the following situations:

» PRF unoccupied in both months – 65 instances.

» PRF occupied in both months – one instance relating to Daubenton's bat.

» PRF occupied in May alone – 11 instances, relating to Bechstein's bat (one account), Daubenton's bat (three accounts), Natterer's bat (three accounts), noctule (one account), common pipistrelle (one account), soprano pipistrelle (one account) and brown long-eared bat (one account).

» PRF occupied in April alone – five instances, relating to Daubenton's bat (three accounts), Natterer's bat (one account), and common pipistrelle (one account).

"Reasonable likelihood" – Maternity groups

To provide a "reasonable likelihood" threshold for maternity groups, Mark Gardener used the Dietz and Pir (2011) account for Bechstein's bat (as set out at Table 11.2) to calculate an encounter probability model that would provide a percentage 'score' for each successive survey.

Gardener demonstrated that in the pregnancy period a single survey visit has a 5% chance of encountering bats if the PRF is occupied by a maternity group. To put it another way, the surveillance team has a 95% probability of being there on the wrong day. Reference to Table 11.2 demonstrates that all species that occupy trees as maternity roosts are to some extent nomadic. Because they are nomadic, two *spaced* visits each have a 5% chance of encountering bats, so in aggregate they *still* have a 95% probability of failing to detect bats, but two *successive* visits have a 9.8% chance of encountering bats. This is because if the tree was empty the first day, the bats may have now moved into the tree on the second. Therefore, the more successive visits that are made, the higher the probability of encountering the colony becomes.

The message is therefore that to increase the probability of finding a suspected 'maternity' roost occupied during the pregnancy period, and thereby being able to identify the bats, count them and sex them (to be sure that they are not in fact a group of males), regularly spaced *sequential visits* will be required if a measure of encounter probability is to be achieved.

Figure 11.3 illustrates the increasing probability of encounter in a situation where an individual PRF was subject to surveillance.

Adopting the threshold of "reasonable likelihood" (i.e. greater than 50% probability), investigating whether an individual PRF is a pregnancy roost might require up to 14 sequential visits in the May/June period.

Unfortunately, missing days is not an option with the colony potentially so mobile. However, the more PRFs that are suitable for an individual species that are surveyed,

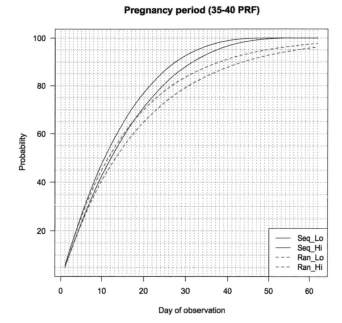

Figure 11.3 Encounter probabilities (as %) during the May/June pregnancy period in which Bechstein's bat maternity colony sub-groups switch roost on average every 1.5 days.

the greater the probability the species will be encountered.[7] Table 11.3 illustrates the number of visits that might be required to demonstrate a *"reasonable likelihood"* of encounter in respect of 1–40 PRFs surveyed on any one day in any one area of wooded habitat.[8]

"Reasonable likelihood" – individuals

Taking the BTHK Database evidence into account, 94% of roosts that were occupied by individual bats in the pregnancy period were occupied in one month alone (65% in May and 29% in June), leaving only 6% occupied in both months. A perfectly reasonable argument might be that at least two visits would be required to have a *"reasonable likelihood"* of detecting a roost occupied by an individual bat at this time. However, in practice the number of visits in the pregnancy period has proven difficult to judge.

If the instances an individual PRF was occupied in both months are added to the individual months, 67% of encounters were recorded in May and 33% were recorded in June, which suggests *"reasonable likelihood"* might be achieved by an individual visit in May.

Proportionality

In a commercial context, it can reasonably be assumed that there is a *"reasonable likelihood"* that 14 visits for an individual PRF would be considered to represent a significant expense!

---

7   In the same way that buying more lottery tickets or betting on more horses increases the probability of winning, the odds only significantly increase if all the tickets are in the same lottery, and the horses are running in the same race.

8   This assumes the trees all hold PRFs that are superficially suitable for the same species in the pregnancy period (which will have been established at the truthing stage), and all occur within a surface area of habitat that is no greater than that exploited by a maternity colony of that species.

Table 11.3 Bechstein's bat maternity colony sub-groups encounter probability (as %) during the May/June pregnancy period where 1–40 PRFs are surveyed on the same day

OBSERVED PRFs	May/June 'Pregnancy period'		
	>50%	80%	95%
1 × PRF	Day 14	Day 31	Day 57
2 × PRF	Day 9	Day 21	Day 38
3 × PRF	Day 7	Day 15	Day 28
4 × PRF	Day 6	Day 12	Day 22
5 × PRF	Day 5	Day 10	Day 18
6 × PRF	Day 4	Day 9	Day 16
7 × PRF	Day 4	Day 8	Day 14
8 × PRF	Day 3	Day 7	Day 12
9 × PRF	Day 3	Day 6	Day 11
10 × PRF	Day 3	Day 5	Day 10
11 × PRF	Day 2	Day 5	Day 9
12 × PRF	Day 2	Day 5	Day 8
13 × PRF	Day 2	Day 4	Day 7
14 × PRF	Day 2	Day 4	Day 7
15 × PRF	Day 2	Day 4	Day 6
16 × PRF	Day 2	Day 3	Day 6
17 × PRF	Day 2	Day 3	Day 5
18 × PRF	Day 2	Day 3	Day 5
19 × PRF	Day 1	Day 3	Day 5
20 × PRF	Day 1	Day 3	Day 4
21 × PRF	Day 1	Day 3	Day 4
22 × PRF	Day 1	Day 2	Day 4
23 × PRF	Day 1	Day 2	Day 4
24 × PRF	Day 1	Day 2	Day 4
25 × PRF	Day 1	Day 2	Day 3
26 × PRF	Day 1	Day 2	Day 3
27 × PRF	Day 1	Day 2	Day 3
28 × PRF	Day 1	Day 2	Day 3
29 × PRF	Day 1	Day 2	Day 3
30 × PRF	Day 1	Day 2	Day 3
31 × PRF	Day 1	Day 1	Day 2
32 × PRF	Day 1	Day 1	Day 2
33 × PRF	Day 1	Day 1	Day 2
34 × PRF	Day 1	Day 1	Day 2
35 × PRF	Day 1	Day 1	Day 2
36 × PRF	Day 1	Day 1	Day 2
37 × PRF	Day 1	Day 1	Day 1
38 × PRF	Day 1	Day 1	Day 1
39 × PRF	Day 1	Day 1	Day 1
40 × PRF	Day 1	Day 1	Day 1

It is therefore worth reiterating items 3 and 4 of the proportionality test set out at Chapter 1:

3. The method and level of effort must be necessary to achieve the aim, that there cannot be any cheaper way of doing it.

4. The measures must be reasonable i.e. suit the circumstances; rarity of the species, conservation trend, etc.

In the case of the barbastelle, Bechstein's bat and Alcathoe's bat, it may be that the full survey programme is reasonable in view of the rarity of the species and the uncertainty of their conservation trends, and the need therefore to establish the numbers present. This is particularly pertinent to the barbastelle, where even the compensation of an individual roost cannot be achieved as maternity groups have not been demonstrated to adopt artificial features. The criteria of the Natural England Licensing Consultation (Natural England 2016) that would allow a lower than usual level of survey effort cannot therefore be satisfied (see Chapter 1). However, it would be wise to weigh the cost of a 14-visit survey, against the cost of trapping and radiotracking. As the cost/benefit margins within some development schemes are insufficient to be able to offer compensation for habitat loss in respect of rare and uncommon bat species, it would also be sensible to discuss what the result would be if a maternity group of such a species were present. If the result would be that the project is abandoned, it may in fact be that radiotracking is not necessary and it can simply be assumed that if a lactating female is trapped a maternity group is present and no further surveillance is required.

Although the May and June period is labelled here as the pregnancy period, in fact Daubenton's bats may begin to give birth in the fourth week of May, Natterer's bat in the first week of June, and brown long-eared bats in the second week of June (D. Linton personal communication April 2016). Referring to Linton's work, as newborn young will be unable to fly and the mothers will have to make returns during their nights foraging to suckle them, it can reasonably be predicted that some evidence of occupancy such as droppings, odour and substrate cues will be apparent even if the colony had come and gone within the 14-day period. Therefore, where close-inspection is feasible, a proportionate approach in respect of a PRF with the potential to be occupied by a maternity group of Daubenton's bats, Natterer's bats, Leisler's bats or brown long-eared bats might be to do two visits in June, at least 15 days apart.[9] This advice would appear to accord with the spirit of the Natural England Licensing Consultation (see Natural England 2016) in respect of common species, but it should be noted that maternity groups of the 'uncommon' Leisler's bat do not appear to adopt artificial roost-boxes, and a greater level of confidence might therefore be required in respect of this species.

For those species that exhibit longer durations of occupancy – Brandt's bat, whiskered bat, noctule and the pipistrelles – a proportionate approach might be to do one visit alone, but leave it until the second half of June to maximise the amount of evidence potentially available. As with the previous paragraph, this advice would appear to accord with the Natural England Licensing Consultation, because noctules readily adopt artificial roost boxes, and maternity groups of the remaining species are encountered in trees so rarely as to be well below the *"reasonable likelihood"* threshold of occurrence from the outset.

For individuals, where surveillance is justified, the data available suggest a proportionate approach might be to do an individual visit in May.

---

9 This might also be augmented by the deployment of a passive ultrasound monitor at the same height as the PRF but on an adjacent tree, in the hope of recording patterns of contacts or dawn-swarming behaviour that might indicate the presence of a maternity group.

11.3.4 Nursery
Number of bat species potentially present
Ten bat species have been recorded occupying trees in the nursery period:
- » The barbastelle.
- » Bechstein's bat.
- » Alcathoe's bat.
- » Daubenton's bat.
- » Natterer's bat.
- » The noctule.
- » Leisler's bat.
- » The common pipistrelle.
- » The soprano pipistrelle.
- » The brown long-eared bat.

Male and female behaviour
During the nursery period, save for that in relation to male young of that year, the tolerance exhibited by the females for the males remains low.

Activity
Reference to Figure 11.2 demonstrates that the nursery period is that in which it is rarest to encounter a torpid bat.[10]

Duration of occupancy
The current knowledge in respect of the duration of occupancy for maternity groups in the nursery period, is broadly defined for all species at Table 11.2.

However, in order to provide some degree of sensible advice in respect of individuals, paired records (i.e. where the same PRF was inspected in both July and August) were identified and the results collated. In total, there are 71 paired records on the BTHK Database, comprising the following situations:
- » PRF unoccupied in both months – 56 instances.
- » PRF occupied in both months – eight instances, relating to barbastelle (two accounts), Bechstein's bat (one account), Daubenton's bat (two accounts), Natterer's bat (two accounts) and soprano pipistrelle (one account).
- » PRF occupied in July alone – five instances, relating to Daubenton's bat (one account), Natterer's bat (two accounts), soprano pipistrelle (one account) and brown long-eared bat (one account).
- » PRF occupied in August alone – two instances, relating to Daubenton's bat (one account) and Natterer's bat (one account).

"Reasonable likelihood" – maternity groups
As with the pregnancy period, to provide a *"reasonable likelihood"* threshold for maternity groups, Gardener used the Dietz and Pir (2011) account for Bechstein's bat (as set out at

---

10 In fact, all the records that relate to torpid bats in the nursery period are of barbastelles, which are unusual in their habit of closing their eyes when they perceive an endoscope. Their pelage also suggests a background-matching strategy and individual bats will typically remain very still in the roost during inspections, where other species typically retreat. It is therefore perfectly possible that the barbastelles encountered during August were in fact awake and merely gave the appearance of torpidity.

Table 11.2) to calculate an encounter probability model that would provide a percentage 'score' for each successive survey. The results are different because as the size of the group increases with recruitment by August, the bats need more space, and the colony is therefore typically divided between three trees rather than two (in the overall resource of around 40). However, these trees are still typically in close proximity (Dietz and Pir 2011).

Figure 11.4 and Figure 11.5 illustrate the increasing probability of encounter in a situation where an individual PRF was subject to surveillance in July and August, respectively.

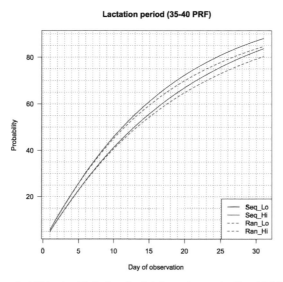

Figure 11.4 Encounter probabilities (as %) during the July lactation period in which Bechstein's bat maternity colony sub-groups switch roost on average every 3.3 days.

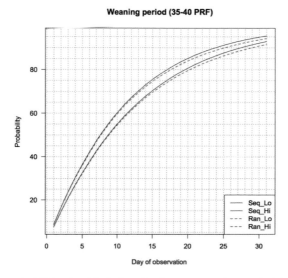

Figure 11.5 Encounter probabilities (as %) during the August weaning period in which Bechstein's bat maternity colony typically divide into three sub-groups and switch roost on average every 3.3 days.

Although the rule with regard to regularly spaced sequential visits still applies, as the bats are now moving at 3.3- and 4.8-day intervals, it is possible to miss every second day in July and every second and third day in August without significantly compromising the survey.[11] Therefore, adopting the threshold of *"reasonable likelihood"* (i.e. greater than 50% probability) suggests that surveillance of an individual PRF might require up to seven visits over 13 days (every other day) in July, and four visits over 10 days (every third day) in August.

As with the pregnancy period, increasing the number of PRFs surveyed increases the probability of encounter. Table 11.4 illustrates the number of visits that might be required to demonstrate a *"reasonable likelihood"* of encounter in respect of 1–40 PRFs surveyed on any one day in any one area of wooded habitat. Note: the same principle in respect of missing days applies in the context of increased sample sizes. Therefore, the number of visits cited in the tables represent the range of days over which the surveillance is performed not the number of actual visits required in that period. For example, if nine PRFs are surveyed in July >50% can be achieved by performing surveillance on two days – Day 1 and Day 3 – with a day off in-between.

"Reasonable likelihood" – individuals

Taking the BTHK Database evidence into account, 46% of roosts that were occupied by individual bats in the nursery period were occupied in one month alone (33% in July and 13% in August), leaving 54% occupied in both months. A perfectly reasonable argument might be that at least two visits would be required to have a *"reasonable likelihood"* of detecting a roost occupied by an individual bat at this time. However, in practice the number of visits in the nursery period has proven difficult to judge.

If the instances an individual PRF was occupied in both months are added to the individual months, 56% of encounters were recorded in July and 44% were recorded in August, which suggest *"reasonable likelihood"* might be achieved by an individual visit in July.

Proportionality

The discussion and suggestions set out in respect of proportionality in the pregnancy period are equally applicable in the nursery period, and the three-tier approach may again be applied:

1. Detailed surveillance in respect of barbastelle, Bechstein's bat and Alcathoe's bat maternity groups.
2. Either end visits; 14 days apart in July or 11 days apart in August, for Daubenton's bat, Natterer's bat, Leisler's bat and brown long-eared bat.
3. An individual visit in July and August for all other species.

For individuals, where surveillance is justified, the data available suggest a proportionate approach might be to do an individual visit in July.

Notes

Reference to Table 11.2 demonstrates that the duration of roost occupancy exhibited by maternity groups increases significantly as the nursery period progresses. This alone has the effect of increasing the probability of an incidental encounter. However, the probability of detecting field signs is also increased by three factors that are particular to this period:

---

11 This is not a guarantee that an individual or a group might have come and gone on the day that was missed (i.e. arrived that morning and decamped the following evening), but it represents an acceptable level of risk.

Table 11.4 Bechstein's bat maternity colony sub-groups encounter probability (as %) during the July and August nursery period where 1–40 PRFs are surveyed on the same day

OBSERVED PRF	July 'Lactation period' INSPECTIONS EVERY OTHER DAY			August 'Weaning period' INSPECTIONS EVERY THIRD DAY		
	>50%	80%	95%	>50%	80%	95%
1 × PRF	Day 14	Day 31	Inf	Day 9	Day 21	Inf
2 × PRF	Day 9	Day 21	Inf	Day 7	Day 16	Day 28
3 × PRF	Day 7	Day 15	Day 28	Day 6	Day 12	Day 22
4 × PRF	Day 6	Day 12	Day 22	Day 5	Day 10	Day 19
5 × PRF	Day 5	Day 10	Day 19	Day 4	Day 9	Day 16
6 × PRF	Day 4	Day 9	Day 16	Day 4	Day 8	Day 14
7 × PRF	Day 4	Day 8	Day 14	Day 3	Day 7	Day 12
8 × PRF	Day 3	Day 7	Day 12	Day 3	Day 6	Day 11
9 × PRF	Day 3	Day 6	Day 11	Day 3	Day 6	Day 10
10 × PRF	Day 3	Day 5	Day 10	Day 2	Day 5	Day 9
11 × PRF	Day 2	Day 5	Day 9	Day 2	Day 5	Day 8
12 × PRF	Day 2	Day 5	Day 8	Day 2	Day 4	Day 7
13 × PRF	Day 2	Day 4	Day 7	Day 2	Day 4	Day 7
14 × PRF	Day 2	Day 4	Day 7	Day 2	Day 4	Day 6
15 × PRF	Day 2	Day 4	Day 6	Day 2	Day 3	Day 6
16 × PRF	Day 2	Day 3	Day 6	Day 2	Day 3	Day 6
17 × PRF	Day 2	Day 3	Day 5	Day 2	Day 3	Day 5
18 × PRF	Day 2	Day 3	Day 5	Day 1	Day 3	Day 5
19 × PRF	Day 1	Day 3	Day 5	Day 1	Day 3	Day 5
20 × PRF	Day 1	Day 3	Day 4	Day 1	Day 3	Day 4
21 × PRF	Day 1	Day 3	Day 4	Day 1	Day 2	Day 4
22 × PRF	Day 1	Day 2	Day 4	Day 1	Day 2	Day 4
23 × PRF	Day 1	Day 2	Day 4	Day 1	Day 2	Day 4
24 × PRF	Day 1	Day 2	Day 4	Day 1	Day 2	Day 3
25 × PRF	Day 1	Day 2	Day 3	Day 1	Day 2	Day 3
26 × PRF	Day 1	Day 2	Day 3	Day 1	Day 2	Day 3
27 × PRF	Day 1	Day 2	Day 3	Day 1	Day 2	Day 3
28 × PRF	Day 1	Day 2	Day 3	Day 1	Day 2	Day 3
29 × PRF	Day 1	Day 2	Day 3	Day 1	Day 2	Day 2
30 × PRF	Day 1	Day 2	Day 3	Day 1	Day 1	Day 2
31 × PRF	Day 1	Day 1	Day 2	Day 1	Day 1	Day 2
32 × PRF	Day 1	Day 1	Day 2	Day 1	Day 1	Day 2
33 × PRF	Day 1	Day 1	Day 2	Day 1	Day 1	Day 2
34 × PRF	Day 1	Day 1	Day 2	Day 1	Day 1	Day 2
35 × PRF	Day 1	Day 1	Day 2	Day 1	Day 1	Day 2
36 × PRF	Day 1	Day 1	Day 2	Day 1	Day 1	Day 1
37 × PRF	Day 1	Day 1	Day 1	Day 1	Day 1	Day 1
38 × PRF	Day 1	Day 1	Day 1	Day 1	Day 1	Day 1
39 × PRF	Day 1	Day 1	Day 1	Day 1	Day 1	Day 1
40 × PRF	Day 1	Day 1	Day 1	Day 1	Day 1	Day 1

The 'Day' refers to the interval and not the number of visits. For example, where one PRF was inspected in July, "reasonable likelihood" would be achieved by seven visits performed every other day, which means the survey would in fact conclude on Day 13. In the same scenario in August, "reasonable likelihood" would be achieved by four visits, performed every three inspections at the 3-day interval and a fourth either on a 2-day interval or completing on Day 10 at the discretion of the survey team.

» Increased duration of occupancy means a greater build-up of droppings even in the absence of babies.

» When babies are born, the group size doubles, and while half the bats present cannot fly, all their droppings will be deposited in the roost.

» The females must now return to the roost at intervals during the night to nurture their young. As a result, rather than going out and coming back once, they may exit and return twice, three times and even more. Substrate cues can therefore be predicted to be conspicuous in roosts occupied by maternity groups.

11.3.5 Mating

Number of bat species potentially present

Eleven bat species have been recorded occupying trees in the mating period:

» The barbastelle.

» Bechstein's bat.

» Alcathoe's bat.

» Daubenton's bat.

» Natterer's bat.

» The noctule.

» Leisler's bat.

» Nathusius' pipistrelle.

» The common pipistrelle.

» The soprano pipistrelle.

» The brown long-eared bat.

Although there is at present insufficient data to test the hypothesis, field experiences perceive a redistribution of species across habitats during the mating season. If this perception is broadly correct, PRFs in more open habitats that had been empty in the earlier part of the year may now be occupied by bat species that are present in that situation at no other time. The same may be true not only of previously unoccupied PRFs in woodland, but also roost features that had been exploited by maternity roosts of one species may now be occupied by another species for mating. For example, a working hypothesis at this time is that there is an association between maternity colonies of brown long-eared bats and mating noctules in that the two exploit the same PRF in different periods.

Male and female behaviour

In the mating season, the intolerance of the females switches to attraction.

The species can be divided into two camps:

» Species that become more conspicuous at tree-roosts:

– The noctule;

– Leisler's bat;

– Nathusius' pipistrelle;

– The common pipistrelle; and

– The soprano pipistrelle.

» Species that do not become more conspicuous at tree-roosts:

– The barbastelle;

- All *Myotis* species;
- The brown long-eared bat; and
- The lesser horseshoe-bat.

During the mating season, male noctules occupy mating-roosts in trees that are defended from other males and from which they call to attract the females (Dietz *et al.* 2011). Calling typically commences 1.5–2 hours following sunset and continues for a significant period. It is suggested that mating takes place in the day (Dietz *et al.* 2011). This indicates that these are roosts that will be occupied continuously both day and night for a relatively long period.

British accounts of the mating display of the male Leisler's bat are lacking. An account from Greece suggests that the males may sing from an exposed roost on the woodland-edge, and lead females by stages to mating roosts deep within the woodland (Dietz *et al.* 2011). Whether these roosts are defended from other males, and whether mating takes place in the day, is unknown.

Reference to Russ (2012) rewards with a tantalising description of what he terms 'advertisement calls'. Russ notes that although the pipistrelles may give advertisement calls at any point over the active year, there is a sudden peak beginning in early September and dropping dramatically in early October. While male common and soprano pipistrelles utter these calls in flight (within a territory of *c.* 200 m), approximately 90% of Nathusius' pipistrelle calls are uttered from the roost itself (Russ 2012).

Ninety-one per cent of common and soprano pipistrelle records on the BTHK Database in the mating period are of individual bats, which might suggest that while the males remain in the trees during the day, the females are only present at night and the sexes spend their days apart. However, this is not supported by accounts of mating in artificial roost-boxes, where mixed-sex groups comprising one male and one or more females are the norm.

Jahelková and Horáček (2011) found that Nathusius' pipistrelle males often grouped together with pairs and trios singing from adjacent trees, so if one mating roost is found, there is a *"reasonable likelihood"* another is present in the vicinity. Furthermore, on the continent at least, display roosts of the Nathusius' pipistrelle have been found in trees in the vicinity of the subterranean roosts in which the females overwinter (M. Van De Sijpe 2015 personal communication – see Andrews *et al.* 2016); the conspicuous mating-roosts of the Nathusius' pipistrelle may therefore lead to the more cryptic hibernacula.

The very fact that the other species are not conspicuous in trees (which is logical given their relatively quiet calls and the complexity of the habitat that would further diffuse the sound) suggests that male barbastelles, *Myotis* spp. or brown long-eared bats do not attempt to advertise themselves to the females from tree-roosts, nor do they attempt to defend individual roosts from rivals.

Save for one instance of two male Daubenton's bats having been netted from a tree-roost in September, the sex of the bats encountered in tree-roost records held on the BTHK Database is unknown. Nevertheless, the records were examined to see whether there was any pattern that was immediately apparent within the data. To this end, records of the barbastelle, Bechstein's bat, Daubenton's bat, Natterer's bat, brown long-eared bat and lesser horseshoe-bat were collated for September and October and divided into two groups: one bat and more than one bat. The results are presented at Table 11.5. In summary, when it is noted that the records of Bechstein's bat groups are biased by continuous observations of two roosts where the numbers were constant, the division between the number of records involving individual bats and groups is not pronounced, and given the wide range in group sizes recorded in September, even developing a hypothesis as

Table 11.5 Records of the barbastelle, Bechstein's bat, Daubenton's bat, Natterer's bat, brown long-eared bat and lesser horseshoe-bat held on the BTHK Database, collated for September and October and divided into two groups: one bat and more than one bat

	SEPTEMBER		OCTOBER	
	1 bat	2+ bats	1 bat	2+ bats
Barbastelle	—	1 record (2 bats)	1 record	1 record (2 bats)
Bechstein's bat	4 records	15 records (3–10 bats)	—	—
Daubenton's bat	4 records	1 record (2 bats; both males)	1 record	—
Natterer's bat	7 records	6 records (2–26 bats)	10 records	1 record (3 bats)
Brown long-eared bat	5 records	7 records (2->10 bats)	13 records	5 records (2–3 bats)
Lesser horseshoe-bat	—	1 record (2 bats)	—	—

to whether the barbastelle, *Myotis* spp., and brown long-eared bats mate in trees is confounded. Two questions that arise are therefore:

» Other than the *Nyctalus* and *Pipistrellus* spp., do the other tree-roosting species mate in trees at all?

» If the *Myotis* spp. and brown long-eared bat, all of which swarm at the entrances to subterranean features (see Parsons *et al.* 2003; Glover and Altringham 2008), also mate there, then where does the barbastelle mate?

Activity
Reference to Figure 11.1 demonstrates that all the bats encountered in trees in September were awake, and 70% of encounters in October relate to bats that were awake and aware of the inspection.

Duration of occupancy
Looking at repeat records held on the BTHK Database, there are an overall 79 paired records where the same PRF was inspected in September and October of the same year, comprising the following situations:

» PRF unoccupied in both months – 60 instances.

» PRF occupied in both months – six instances: common pipistrelle (one account), brown long-eared bat (one account), and four changes of species (one account of a barbastelle being replaced by a Natterer's bat, one account of a soprano pipistrelle being replaced by a Natterer's bat, and two accounts of brown long-eared bats being replaced by noctules).

» PRF occupied in September alone – ten instances: Bechstein's bat (two accounts), Natterer's bat (three accounts), noctule (two accounts), soprano pipistrelle (one account), and brown long-eared bats (two accounts).

» PRF occupied in October alone – three instances: Natterer's bat (one account), and noctule (two accounts).

Despite the suggestions of territoriality during the mating period, the pipistrelles appear notably absent from the BTHK Database, and the noctules still appear mobile. This appears to accord with the findings of Hopkirk and Russ (2004) in respect of Leisler's bat, where in their pre-hibernal and hibernal study they recorded a peak in roost-switching in the last week of September and again between the end of September and early November, with the highest level of switching taking place in the first week of October.

"Reasonable likelihood"

Taking the BTHK Database evidence into account, 67% of roosts that were occupied in the mating period, were occupied in one month alone (50% in September and 17% in October), leaving 33% occupied in both months (and not always by the same species). As with the spring-flux period, a perfectly reasonable argument might be that at least two visits would be required to have a *"reasonable likelihood"* of detecting a roost in this period.

However, if the instances an individual PRF was occupied in both months are added to the individual months, 62.5% of encounters were recorded in September and 37.5% were recorded in October, which sits comfortably with Jon Russ's description of the greatest intensity of 'advertisement calling' (Russ 2012). The data, and the accounts of encounters of mating behaviour given by other authors, suggest *"reasonable likelihood"* might be achieved by an individual visit in in the period spanning the second through last week of September.

Proportionality

Where surveillance is justified, the data available suggest a proportionate approach might be to do an individual visit in mid- to late-September.

11.3.6 Autumn-flux

Number of bat species potentially present

Eleven bat species have been recorded occupying trees in the autumn-flux period:

- » The barbastelle.
- » Bechstein's bat.
- » Alcathoe's bat.
- » Daubenton's bat.
- » Natterer's bat.
- » The noctule.
- » Leisler's bat.
- » Nathusius' pipistrelle.
- » The common pipistrelle.
- » The soprano pipistrelle.
- » The brown long-eared bat.

Male and female behaviour

The bats that have been recorded in trees in the autumn-flux period have not been sexed, and it is therefore unknown whether there is any bias generally, or within any species. However, as with the winter period, the general trend appears to be toward tolerance between the sexes at this time.

Activity

There is a perceived tailing-off in activity during the spring-flux, but the data presented in Figure 11.1 is skewed by one species, the brown long-eared bat, which comprises 56%

of the records. While the number of bats that are torpid and those that are awake is broadly similar, 83% of brown long-eared bats encountered at this time were asleep if not fully torpid. The barbastelle records relate to an even split. The Daubenton's and Natterer's bats encountered were for most part awake, and all the noctules and pipistrelles encountered were awake.

Duration of occupancy
Looking at repeat records held on the BTHK Database, there are an overall 145 paired records where the same PRF was inspected in November and December of the same year, comprising the following situations:

» PRF unoccupied in both months – 116 instances.

» PRF occupied in both months – 19 instances: Natterer's bat (two accounts), noctule (three accounts), common pipistrelle (two accounts), soprano pipistrelle (one account), brown long-eared bat (eight accounts), and three changes of species (one account of a barbastelle being replaced by a common pipistrelle, one account of a Natterer's bat being replaced by a common pipistrelle, and one account of a Natterer's bat being replaced by a brown long-eared bat).

» PRF occupied in November alone – seven instances: Daubenton's bat (one account), Natterer's bat (three accounts), and brown long-eared bats (three accounts).

» PRF occupied in December alone – three instances: Natterer's bat (one account), common pipistrelle (one account) and brown long-eared bat (one account).

"Reasonable likelihood"
Taking the BTHK Database evidence into account, 66% of roosts that were occupied during the autumn-flux period were occupied in both months. Even if the instances when the species had changed are removed and added to individual months, the proportion of roosts that were occupied by the same species in both months is 55% and therefore still crosses the 'over 50%' threshold of *"reasonable likelihood"* at 55%, and might therefore suggest only an individual visit is required. By placing all the encounters with roosting bats into their individual months, 54% were in November and 46% were in December, suggesting that November is the marginally better month to target.

Therefore, despite many species still being more awake than torpid, the perception is of long-duration occupancies. As a result, the limited data available do support the suggestion that a well-timed visit from the second half of November would appear to have a *"reasonable likelihood"* of assessing the status of a PRF as a roost.

Proportionality
Where surveillance is justified, the data available suggest a proportionate approach might be to do an individual visit in mid- to late-November.

Notes
Blue tits are very much in evidence night-roosting in trees during the autumn-flux period. As a result, in the absence of any other evidence, substrate cues might be tested by an evening visit to make sure they are not attributable to a bird.

CHAPTER 12

Trouble-Shooting

In this chapter	
Introduction	Uncertainty
Fundamental principles	Matching your approach to a client's expectations
Inconclusive and 'null' results	A catch-all due-diligence safeguarding strategy
Indirect damage and disturbance	The need to consider indirect as well as direct effects upon bats occupying tree roosts potentially resulting from development proposals or management operations
Comparative assessment of environment	Environmental Impact Assessment
Late commissions	A lesson in integrity
Unlicensed assessments	What you are allowed to do without a survey licence
Getting a second-opinion	Where to go for advice
Fault-finding	What to do if you find an omission or an error within this book

12.1 Introduction

This chapter is all about uncertainty.

The BTHK project was born of uncertainty, and it is therefore fitting that the book should conclude with a discussion about it.

The following text is taken from conversations with a wide range of people, including frustrated and confounded developers, land-owners, arborists and foresters. Most of the discussion therefore relates to professional scenarios, but the section about what can be done without a licence relates equally to amateur naturalists and also junior surveyors who want to improve their own knowledge and skills.

While not every situation involving uncertainty can be anticipated in a single chapter, the primary reasons for uncertainty and where it might lead to conflict are discussed, and some advice that has been found to be effective in reducing tension and controlling the stress-levels of all parties involved is provided.

12.2 Fundamental principles

In the greater proportion of trouble-shooting scenarios, stripping the issues back to the fundamental principles will simplify the situation by focusing attention onto only those matters that are pertinent in each context.

Fundamentally, in a crisis, the reason an ecological advisor is called in is due to *uncertainty*.

What most clients are seeking is clarification, delivered in a confident manner. In this context, the clarification will typically comprise the identification of the threshold of risk associated with the specific operation that is proposed.

When the initial approach is made it is important to react to questions with questions rather than answers.

Firstly, and this cannot be overemphasized: **if they email you, then you should phone them!**

A rapport is vital if confidence is to be instilled, and this cannot be achieved via email.

When you speak to them, ask them why they have approached *you* specifically. If they picked your name from a web-search, then it is up to you to instil a sense of confidence in them before you get into the reason they called. It helps to give them some background as to who you are, and what it is you do. If they have phoned on a recommendation, it is wise to explain to them what your relationship is with the person who recommended you.

After the initial pleasantries, get them to explain *exactly* why they think they need ecological advice. Then, ask them for a step-by-step explanation of what it is their operation comprises, and how the operation would be performed if there was no potential for bats to be present. After that, identify which step has the potential for legislative conflict in terms of:

» Disturbance to bats.

» Injury or mortality to bats.

» Damage or destruction of a bat roost.

When you have all this information, give them a description of where you think the potential for conflict is, and ask them if they would agree. Make notes of their answers.

Obviously, as ecologists we cannot give professional legal advice, but we cannot give professional ecological advice unless we have all the facts. Ask as many questions as are necessary to give a robust response to the following questions, bearing in mind that if there is later a prosecution all this evidence will be required by their defending counsel. Make this very clear to them. The following is a structured list of questions that might be asked to ensure all the relevant information is collected for review.

» **Question 1:** Why do they think they need professional ecological advice?

» **Question 2:** Have any third-party had involvement? Specifically, establish whether there has been any prior ecological survey or assessment, or a Tree Hazard Assessment. If there has, get copies of the reports.

» **Question 3:** So that you have all the facts and understand the situation, can you have a step-by-step description of how the operation would be performed if there was no suggestion that roosting bats might be present?

» **Question 4:** Might the operation result in any of the following upon roosting bats if they were present:

  – A vibration effect?

  – A noise effect?

  – An increase in light-levels?

  – A perceptible change in the surrounding environment?

  – Any other change they can think of?

  – Can they think of a way that each potential effect might be removed or reduced?

» **Question 5:** Might the operation have the potential for any of the following upon roosting bats if they are present:

- The bat(s) being trapped inside?
- The bat(s) being prevented from getting into a roost?
- Can they think of a way that each potential effect might be removed or reduced?

» **Question 6:** Do they themselves consider that if a bat-roost was present, the operation might result in:

- The roost being damaged?
- The roost being destroyed?
- Can they think of a way that each potential for damage or destruction might be removed or reduced?

Each time you ask a question listen to:

1. What they are telling you.
2. **What they are not telling you.**

If they are considering an operation, but have not yet begun it, they will typically give you all the facts and will provide as complete account as they can.

If they have already begun an operation and then become uncertain, they may give you a carefully thought-out script that is a mixture of fact and fiction. Regardless, it will in most cases be an incomplete account.

Summing-up
When you have their answers, read them back to ensure they agree the account you have is correct.

Both of you now understand what the client is attempting to achieve, and why they have become uncertain as to how best to proceed. The result should be that the client now feels they have shared the problem with someone who is taking their concern seriously.

When a consultant is called out, the person that called them is looking for a *facilitator* – a professional who will make an action or process easier. That process does not mean that the consultant cannot say 'no', but it does mean that they should be able to explain robustly 'why not' in language the client can understand, and conspicuously attempt to provide the client with an alternative course of action with a clearly identified conclusion that is in some way helping (i.e. if you do this, you will get that).

It is helpful to make it obvious that you are making every attempt to help them. This is rather like the exaggerated turning of the head to demonstrate to a driving-examiner you are checking the rear-view mirror; the client needs to really see that an effort is being made to find a solution.

The last thing a drowning man needs is someone describing the water, so when speaking to a client, only mention the law where it is relevant; you can send the detail in an email later.

It is vital to bear in mind that the client has contracted a 'crusader', and any hint that the person who has arrived is a 'conservationist' first and a 'consultant' second will result in conflict. The advice must not be biased by an unsupported conservation agenda, but be cold and objective, identifying where the project might come into conflict with legislative and/or policy mechanisms, and searching loudly for a safeguarding strategy that achieves as much of the client's desired outcome as is possible (even if it is only to deliver part of the desired outcome, and even then, after a delay).

Being grounded in objectivity will limit the potential for conflict.

If the outcome of your discussion is that you consider further action is required in order to provide a due-diligence safeguard, then turn to the start of this book and explain the process involved. If you do this at the outset they will understand what you are there to do, before you set foot on the site.

Finally, it is always better to give bad news early, rather than give the impression that the cake can be had and eaten in its entirety only to have to take slices away at a later date. Give the worst-case scenario over the phone before you go out to the site, with the proviso that it is one potential outcome in a range of options. This is *vital* because the most common cause of trouble after an advisor has been brought in is the failure of that advisor to meet the client's expectations after having painted an overly optimistic picture. Giving them the worst-case scenario over the phone ensures that if they do contract you, they have done so knowing full well what your approach will be. It also gives them time to digest what you have said before you are face to face.

However, and again this cannot be overemphasized: **if any reference was made to a third-party's involvement in answer to Question 2, such as a previous survey, or a Tree Hazard Assessment, make sure you get a copy of the report and review it prior to offering any formal advice.**

After you hang-up, put everything you have discussed in an email so you have a record of the enquiry.

12.3 Inconclusive and 'null' results

Where an appropriate level of surveillance has been performed and the result has been inconclusive, it may be appropriate to adopt a due-diligence safeguarding strategy in order to guard against the potential for legislative conflict. The text below is an example of such a strategy, which has been found to be useful within the context of an overarching Ecological Management Plan.

12.3.1 Legislation: roosting bats

All bat species and their roosts receive legal protection under the *Wildlife & Countryside Act 1981* and the *Conservation of Habitats and Species Regulations 2017*.

Part 1, Section 9, subsection (4)(b & c) of the *Wildlife & Countryside Act 1981* states:

Subject to the provisions of this Part, a person is guilty of an offence if intentionally or recklessly –
> (b) *he disturbs any such animal while it is occupying a structure or place which it uses for shelter or protection; or*
> (c) *he obstructs access to any structure or place which any such animal uses for shelter or protection.*

Part 3, regulation 41, paragraph (1) of the *Conservation of Habitats and Species Regulations 2017* states that:

A person who
> (a) *deliberately captures, injures or kills any wild animal of a European protected species,*
> (b) *deliberately disturbs wild animals of any such species, or*
> (c) *damages or destroys a breeding site or resting place of such an animal,*
> *is guilty of an offence.*

In practical terms it is often helpful to separate intentional, deliberate and reckless actions from the one absolute offence set out at item (d) of the *Conservation of Habitats and Species Regulations 2017*:

> *a person who damages or destroys a breeding site or resting place of* [a bat] *is guilty of an offence.*

> **Note:** The offence of damaging or destroying a breeding site or resting place is an **absolute offence** that does not require any fault elements to be proved to establish guilt.

12.3.2 Pre-development site interest: roosting bats
A bat survey of PRFs in trees in [*insert site name*] resulted in an inconclusive result. Weathering of mature trees within a site may result in the formation of suitable bat-roost features in a relatively short period of time (even a single night). Once such a feature does form, bats may immediately exploit it. Conversely, the same actions may result in the degradation and loss of existing features meaning that the bats that had exploited them have to seek alternative roost sites. This has the effect of making the presence of bats in a site, and their location therein, unpredictable from one year to the next. Therefore, the potential for legislative conflict will be anticipated within the following due-diligence safeguarding strategy:

Stage 1
Prior to any works that may affect trees or wooded habitat generally, the extent of the works will be clearly marked on a plan by the site operator, and a copy of that plan with a text description of the works should be provided to a Licensed Ecologist[1] for review and consultation.

If it is the conclusion of the Licensed Ecologist that there is the potential for damage to a bat roost or disturbance to roosting bats, the safeguarding strategy will proceed to Stage 2.

Stage 2
All trees will either be inspected from the ground or ascended by the Appointed Ecologist during the appropriate period for that viewpoint, in order to assess whether they hold PRFs. If no such features are present, then no further action will be necessary. If, however, PRFs are present, then the safeguarding strategy will proceed to Stage 3.

Stage 3
All trees holding PRFs will be subject to a climb-and-inspect survey or, if this is **impossible**, *using an appropriate alternative method, by a Licensed Ecologist. If no bat roosts are present, the tree(s) may be felled, or remedial safety works performed, without delay. If, however, a bat-roost is found, the safeguarding strategy will proceed to Stage 4.*

Stage 4
A European Protected Species Licence may be required from Natural England in order to close the roost and allow works to proceed within the legislation. This situation, or the potential compensation that might be required, cannot, however, be predicted in advance of the survey.

---

1 For the purposes of this due-diligence safeguarding strategy, '*Licensed Ecologist*' means an ecologist holding a survey licence issued by Natural England to take and disturb bats for the purposes of science and education who has *demonstrable experience* (supported by documented evidence) of searching for, *finding*, identifying and *managing* the species concerned in respect of all life stages, in all seasons and within the relevant habitat context.

12.4 Indirect damage and disturbance

> **Note:** BTHK has no legal specialism and is in no way legally qualified. If doubt exists as to the legality of actions, the Discretionary Advice Service of the appropriate government agency should be consulted and/or independent legal counsel should be sought.

The legal protection given to bats under the *Conservation of Habitats and Species Regulations 2017* can be broadly broken down into that which protects the bats themselves and that which protects their habitat.

At first sight, this may appear straightforward; the bats themselves are legally protected against being killed, injured and disturbed, and their roosts are protected against damage or destruction. However, although commuting routes and foraging habitat are not specifically protected, the judgment of the Supreme Court in the case of *Morge (FC) (Appellant) v Hampshire County Council (Respondent): Hilary Term [2011] UKSC 2 On appeal from: 2010 EWCA Civ 608*[2] was that the legislation afforded to bats with regard to disturbance does give some degree of protection to wider habitat in some situations.

Under Article 12(1)(b) of Council Directive 92/43/EEC on the conservation of natural habitats and of wild fauna and flora (hereinafter the Habitats Directive), the UK is obligated to maintain bat populations at favourable conservation status. Article 1(i) of the Habitats Directive defines *'conservation status of a species'* to mean *'the sum of the influences acting on the species concerned that may affect the long-term distribution and abundance of its populations'*. Furthermore, Article 12(1)(b) provides that: *'Member States shall take the requisite measures to establish a system of strict protection for the* [bat species] *listed in their natural range, prohibiting … (b) deliberate disturbance of these species, particularly during the period of breeding, rearing, hibernation and migration.'*

Deferring to the *Guidance Document on the Strict Protection of Animal Species of Community Interest under the Habitats Directive 92/43/EEC* (European Commission 2007), *Morge v Hampshire County Council* acknowledged that *'Disturbance is detrimental for a protected species e.g. by reducing survival chances, breeding success or reproductive ability.'* This is defined as *'any disturbing activity that affects the survival chances, the breeding success or the reproductive ability of a protected species or leads to a reduction in the occupied area'*.

In defining 'disturbance', the Supreme Court therefore accepted that it encompassed not just situations where the bats were harassed by an intrusive change in their immediate environment (such as a bright light being shone on them or loud noise, etc.), but also might also include changes in the wider environment that disturbed their habitual routine.

In ecological terms, such a situation might result from the severance of a linear landscape element used for migration/commuting to and from a roost, resulting in a fragmentation effect, and/or the loss of foraging habitat that formed part of the territory of a colony thereby rendering the territory unviable.

Taking the disturbance legislation in isolation, Part 3, regulation 41, paragraph (1), bullet (b) of the *Conservation of Habitats and Species Regulations 2017* states:

A person who deliberately disturbs wild animals of any such species is guilty of an offence.

The Supreme Court established definitions for *'deliberate'* and *'disturbance'*, which are set out below. In addition, the word *'population'* is also defined with a specific definition in relation to species of bats.

---

2   For the full transcript of the Morge judgment simply type 'uksc-2010-0120-judgment' into Google and the page will pop up. You can then download the full judgment and press summary as a PDF.

12.4.1 Deliberate

The definition of 'deliberate' is set out in paragraph 33 of *Guidance Document on the strict Protection of Animal Species of Community Interest under the Habitats Directive 92/43/EEC* (European Commission 2007) which states:

> *Deliberate actions are to be understood as actions by a person who knows, in light of the relevant legislation that applies to the species involved, and the general information delivered to the public, that his action will most likely lead to an offence against the species, but intends this offence or, if not, consciously accepts the foreseeable results of his action.*

12.4.2 Disturbance

Disturbance was discussed in the context of a temporary or permanent situation that was unsettling, but at a magnitude greater than simply inconvenience. Looking at disturbance in the context of commuting, a human example of a temporary disturbance might be a road-block that meant a journey to work was so long that a parent could not get their children to school and still be at work on time or return to feed them at the appropriate hour. An example of a permanently disturbing situation might be an obstruction that meant a significant detour was required that made travel costs so expensive that it was no longer worth doing a job, consequently resulting in significant changes in lifestyle. In the context of foraging habitat, a human example might be the closure of a favoured supermarket resulting in increased footfall in the remaining supermarkets, with issues of overcrowding, stock shortfall and car-parking. Any of these situations might be predicted to result in displacement (i.e. an individual or family moving to another area).

Deferring again to the European Commission (2007), it was concluded that the disturbance offence is not limited to significant disturbances of significant groups of animals, but covers all disturbance of EPS. However, although the disturbance does not have to be significant, it must be certain (i.e. specific), identifiable and real, and not fanciful.

The Supreme Court highlighted that a disturbing activity that affects bats *'during the period of breeding, rearing, hibernation and migration is more likely to have a sufficient negative impact on the species to constitute prohibited "disturbance" than activity at other times'*.

Furthermore, it must have a detrimental impact so as to affect the favourable conservation status of the species at *population level*. The Supreme Court cited Article 1 of the Habitats Directive in defining favourable conservation status as a situation where *'the natural range of the species is neither being reduced nor is likely to be reduced for the foreseeable future, and there is, and will probably continue to be, a sufficiently large habitat to maintain its populations on a long-term basis'*.

This suggests that some degree of disturbance in terms of commuting route severance and foraging habitat loss might be acceptable before it constituted an offence. The Supreme Court set out that the proper approach should be to give consideration to the effect on the conservation status of the species at population level and biogeographic level. It is stated that *'the impact must be certain or real, it must be negative or adverse to the bats and it will be likely to be detrimental when it negatively or adversely affects the conservation status of the species'*.

Taking the above into account, it is logical to suppose that an Impact Assessment in respect of a bat roost should therefore establish whether a development proposal might have the potential to result in a *disturbance* sufficient to affect:

» The survival chances;

» Breeding success;

» Reproductive ability, and thereby; and

» The range occupied by a population of each species.

This can only be established through a comparative assessment of the *environment* in which roost trees are in prior to the operation proposed, and the environment in which they will be in during and following the operation proposed. This leads us into the next section.

12.5 Comparative assessment of environment (i.e. environmental impact assessment)

The use of built structures by individual species of bats is well documented, but the published accounts lack detail, and our understanding of which species use which structures, how, when and where (within the fabric) is still only patchily understood.

In a built structure, different bat species occupy different niches in terms of exposure and height (ranging from under ridge tiles, slates or felt, behind barges, into the soffit box, cavity wall, or roof void, against the ridge board, into the mortise joints and finally down into the basement). It is now widely understood that taking off the roof, even only in part and for a short duration, will almost certainly disturb any bats present, and this effect will differ between species and roost purpose, depending upon the timing and duration of the works.

The use of subterranean habitats by individual species of bats is well documented, but poorly understood.

Different subterranean habitats have different sizes and designs. In a cave system, different bat species occupy different niches in terms of exposure and depth. It is now widely understood that blocking-up or altering a cave entrance (even by grilling) has not only the potential to destroy the PRF, but can also significantly alter the environmental conditions over a wide area of the remaining habitat.

The use of wooded habitats by individual species of bats is well documented, but poorly understood – that is what the BTHK Project is trying to correct.

What we do know is that all wooded habitats are worth more than the sum of their parts. No wooded habitat functions as an aggregation of individual trees, any more than a house functions as an aggregation of individual rooms, or subterranean habitats function as an aggregation of individual chambers. The overall environment is dependent upon all the individual components: remove one, and risk an impact (the effect of which typically cannot be predicted).

Broadly speaking, wooded habitats, houses and subterranean habitats all offer a gradation of exposed PRFs into cryptic PRFs, with increased levels of buffering from wind, temperature, humidity and light[3] the further from the edge the PRF is situated. The situation in respect of the potential for alterations in environment to have a catastrophic effect upon the functionality of a roost is no different in a situation involving a house, a cave, a wood, or even an individual tree.

What does differ between the three habitats is the reversibility of the environmental impact.

In the case of both the built structure and the subterranean situation, the change in the environment can be mitigated within the space of a few days, and this interval can be shortened still further by an increase in manpower. In contrast, while chopping down trees with the compensation of replanting might be temporary in the life of an elephant, it will represent permanent change in the lifetime of a bat. Even felling a corridor through a wood will have a significant effect upon the environment, and this will be likely to span several generations of the same bat colony. Any reduction of the surface area represents

---

3   In general terms, the more exposed PRF in all three situations tend to be crevices, and the most cryptic tend to be voids.

the loss of an existing and *future* PRF resource, and alters the environment by introducing, enlarging or moving the edge at the expense of the interior.

As a result, to understand the magnitude of an impact and what its effect is likely to be, surveys of wooded habitat that are performed in support of Planning Applications may need to encompass *environmental surveillance* as well as bat surveillance.

In these situations, the surveillance might seek to compare the situation currently present in the Zone of Influence against a control. For example, if an operation was proposed that comprised the felling of a corridor through a woodland, the environment on the line of the corridor could be compared with a pre-existing corridor (a ride, rack, or bridleway, etc.) that was as close in character to the corridor proposed (width, alignment, etc.). Ideally, temperature, humidity, wind speed and lux-levels would be taken on both alignments and at a buffer either side to see over what area the operation might have an effect. If compensation is proposed, there might also be some attempt to predict the duration of the effect in terms of how long it will take for the trees to grow using an evidence base and putting it in the context of the average lifetime of the species that might be affected, and the number of generations that will pass in the interval between the operation and a return to the pre-existing situation. Bat surveillance alone will not inform an assessment of effects; the *combined results* of bat and environmental surveillance will be required.

In a situation involving a colony of a rare or uncommon tree-roosting bat species, with an unknown or declining population trend, such an Impact Assessment will be vital if the likely effect of the operation proposed is to be robustly identified and completely understood.

Objective questions at this stage might include:

1. *Is it 'more-likely-than-not' that the operation will result in an alteration in the temperature or humidity range currently present around an existing roost?*
2. *Is it 'more-likely-than-not' that the operation will result in an alteration in the amount of light reaching an existing roost?*
3. *Is it 'more-likely-than-not' that the operation will result in an alteration in the effect of wind upon an existing roost?*
4. *Is it 'more-likely-than-not' that the operation will result in noise or vibration that is perceptible to bats occupying an existing roost?*
5. *Is it 'more-likely-than-not' that the operation will have an isolation effect by severing or obstructing a flight-line?*

And what data do we need to:

1. Determine what the effect will be.
2. When the effect will occur.
3. How long the effect will last.
4. Predict the magnitude of the impact.
5. Design a strategy to prevent the impact from having the effect that the natural range of a rare or uncommon bat species is reduced?

Finally, how will we monitor the effects? What will be the trigger for action, and what surveillance will be required to ensure that if the trigger threshold is exceeded remedial action will be immediate?

12.6 Late commissions

Every year someone is approached and asked to begin a roost-survey of woodland in the summer months, and every year they approach BTHK for advice.

The BTHK advice is always the same: be honest, can you recommend a course of action that will result in the same standard of surveillance that would have been achieved if the mapping had been done in the winter? If you can, then there is not a problem. If you cannot, you have your answer.

Some clients will tell you:

» **Their project is a one-off …**
  – It is not, they have lots more where this one came from.

» **They are a special case …**
  – They are not, there are lots of them out there.

» **That you have to understand their position …**
  – You do not, you have to be objective.

» **That you are being unreasonable if you do not do exactly what they want, when they want …**
  – You are not, you are being professional.

No client ever dreamt up a development for the good of mankind and it would be a rare situation indeed where a development had to go through urgently because otherwise people would die.

If the only sticking-point is money, then what the developer is asking is for you to compromise yourself for their enrichment. If you acquiesce and perform a survey of a low standard, they may boast about how they managed to bully you into a compromise in their favour, and your reputation will be at risk. If your survey is then questioned, they will nonetheless complain and this will make you unhappy. Regardless, these people typically prove to be high maintenance.

It is the job of a Local Authority Ecologist to question every survey report put in front of them, and they do this even when they are given a solid gold, Swiss-made survey report with jewelled movements. What do you think they will do if they are handed a plastic cuckoo-clock?

What if the application goes to Public Inquiry? It is astonishing how many developers appear to be sufficiently belligerent that they are prepared to pay legal costs when they quibble every survey cost. If you go to Inquiry (even if you did a good job), the opposite side's barrister will do their utmost to put your competence and integrity in doubt. Need we say more?

Unless you have a strong desire for the greater part of your professional vocabulary to be *'Do you want to go large on that?'* stand firm, and live to fight another day with your integrity intact.

It is far more fun to listen to their flannel and then tell them with a smile that, sadly, you are fully booked and cannot meet their timescale, and then recommend a competitor you really do not like, sure in the knowledge that they will make each other thoroughly miserable.

12.7 Unlicensed assessments

The legislation protecting all bat species discourages a good deal of amateur naturalists from looking for tree-roosts, because inspection results in disturbance which is legally prohibited.

In fact, it also puts off people who work with trees (even in a voluntary capacity), and may even act as a constraint to people who are in the early stages of working with bats.

The legislation that prohibits disturbance is the *Wildlife & Countryside Act 1981*, which states:

> Subject to the provisions of this Part, a person is guilty of an offence if intentionally or recklessly –
>> (f) he disturbs any such animal while it is occupying a structure or place which it uses for shelter or protection.

When this legislation came into force, the disturbance offence meant that in order to survey for bats, and indeed any of the species that were now protected, some sort of 'get out of jail free card' had to be created by the Department for Environment, Food & Rural Affairs (DEFRA), and essentially this is what the survey licence is. However, the story does not end here; a little over a decade later *The Conservation (Natural Habitats, &c.) Regulations 1994* also came into force and by stages was revised as the various aspects have been tested and more exactly defined, with the most recent revisions evolving into the *Conservation of Habitats and Species Regulations 2017*.

The legislation that guards against disturbance in the Habitats Regulations is weaker than that in the *Wildlife & Countryside Act 1981*, but the legislation that guards against roost destruction is stronger, and states:

> A person who damages or destroys a breeding site or resting place of [a bat] is guilty of an offence.

Damaging or destroying a bat roost is therefore an absolute offence that does not require any fault elements to be proved to establish guilt; essentially, if you cut down a tree holding a bat roost, you are guilty of a crime.

This leaves unlicensed arborists, foresters and even conservation volunteers with a dilemma: you may be guilty of an offence under the *Wildlife & Countryside Act 1981* if you search for a bat in a tree and inadvertently disturb it, but you definitely will be guilty of an offence under the Habitats Regulations if you do not search and then accidentally damage or destroy a roost.

Arborists, foresters and conservation volunteers therefore have a choice: either risk going to jail under the *Wildlife & Countryside Act 1981* or risk going to jail under the *Habitats Regulations*, which is of course a Catch-22.

This all came to a head a decade after the *Habitats Regulations* came into force, when a new method for establishing dormouse *Muscardinus avellanarius* presence using nest-tubes was published by Chanin and Woods (2003). Nest-tubes were cheap, lightweight, easy to install and simple to check, but dormice had the same level of legal protection afforded to bats, and very few people had a dormouse survey licence, so the uptake was relatively low. As a result, dormouse habitat was still being destroyed due to ignorance.

In response to the Catch-22, English Nature (now Natural England) decided to take a pragmatic approach, and in the 2nd edition of the *Dormouse Conservation Handbook* (Bright *et al.* 2006) they included this paragraph:

> Inspecting nest boxes (and nest tubes) requires a licence from English Nature or the Country-side Council for Wales in areas where dormice are already known to be present, BUT if boxes

or tubes are put out speculatively to detect presence, this in itself does not require a licence, but a licence is essential once the first dormouse has been found.

In summary, what this means is that if you know dormice are present, you should not be messing them about without a licence, but if you do not know whether they are present, and nor does anyone else, then you can have a look to see if they are. This allows anyone to go out and look for dormice whether they have a licence or not, on the understanding that as soon as the species is encountered the unlicensed naturalist will withdraw and seek qualified licensed advice.

Finally, nearly a decade later, the British Standards Institute published the following:

» British Standards Institute 2015. *BS 8956 – Surveying for Bats in Trees and Woodland.* London: BSI.

BS 8596 set out that non-specialist assessments of PRFs could be undertaken by anyone involved in tree management who has had basic bat awareness training, but if a roost is found, the surveyor (if unlicensed) should retreat immediately and seek qualified licensed advice. The Standard specifically refers to:

» The close examination of the PRF to assess its suitability to hold roosting bats.

» The use of a high-powered torch to look in the PRF.

Not having a licence does not therefore mean that you should not look. It simply means you should not interfere with a known roost. The policy now is that arborists, foresters and conservation volunteers should be looking every time tree-works are being considered. Reading this book might reasonably be regarded as basic bat awareness training, and if you read the whole thing from cover to cover, by the end of it you should know enough about tree-roosting ecology to look for roosts in order to satisfy your own curiosity.

12.8 Getting a second opinion

Everyone encounters situations where they are not entirely confident as to how they might proceed with a surveillance programme, a mitigation strategy or the design of compensatory roost provision.

No one has all the answers and, even where a course of action is broadly defined, it never hurts to get a second opinion.

One of the objectives of the BTHK project is to get people to share knowledge.

Social networking is a vital part of our development, and websites such as Facebook allow uncertainty to be shared with the wider bat- and tree-working community. Messages to the BTHK Facebook page can be answered discretely or posted onto the page for wider discussion. In addition, joining groups such as Tree Climbing Bat Surveys and UK Bat Workers means that problems that are encountered regularly are aired, and solutions are shared.

Do not sit stewing about a problem: ask your peers. If no one has the answer, you do at least know no one else could have done the job better!

12.9 Fault-finding

Every attempt has been made to ensure this book is factual and complete with each course of action culminating in a conclusion with evidence-supported advice.

Nevertheless, you are encouraged to actively attempt to attack every part of this book. If you can find any loose-end that leaves the reader with an uncertain conclusion, we need to know where that failing lies so the situation can be corrected in a subsequent edition. If you can find an error, we need to know about it immediately so it can be corrected in a supplement.

PRF Summary Tables

Broadleaved tree species

Disease & Decay PRF

TREE SPECIES	DISEASE & DECAY PRF THE BROADLEAVED TREE SPECIES FORMS THAT ARE KNOWN TO BE EXPLOITED BY ROOSTING BATS (AS DEMONSTRATED BY RECORDS WITH ACCOMPANYING PHOTOGRAPHS ON THE BTHK DATABASE OR PHOTOGRAPHIC ACCOUNTS WITHIN BAT TREE HABITAT KEY (Andrews *et al.* 2016))									MINIMUM DBH OF DISEASE & DECAY ROOST TREE RECORDS FOR THE TREE SPECIES HELD ON DATABASE
	Woodpecker-holes	Squirrel-holes	Knot-holes	Pruning-cuts	Tear-outs	Wounds	Cankers	Compression-forks	Butt-rots	
Field maple *Acer campestre*	—	—	✓	—	✓	✓	—	—	—	7 cm
Norway maple *Acer platanoides*	—	—	✓	—	—	—	—	—	—	99 cm
Sycamore *Acer pseudoplatanus*	—	—	—	—	✓	✓	✓	—	—	9 cm
Horse chestnut *Aesculus hippocastanum*	—	—	—	—	✓	—	✓	—	—	50 cm
Alder *Alnus glutinosa*	✓	—	—	—	✓	✓	—	—	—	15 cm
Silver birch *Betula pendula*	—	—	✓	—	✓	—	✓	—	—	35 cm
Downy birch *Betula pubescens*	—	—	—	—	✓	—	—	—	—	30 cm
Sweet chestnut *Castanea sativa*	—	—	—	—	—	✓	—	—	—	37 cm
Hazel *Corylus avellana*	—	—	—	—	—	✓	—	—	—	10 cm
Beech *Fagus sylvatica*	—	—	✓	---	—	✓	✓	✓	✓	20 cm
Ash *Fraxinus excelsior*	✓	✓	✓	✓	✓	✓	✓	—	—	12 cm
Sweetgum *Liquidambar styraciflua*	—	—	—	—	✓	—	—	—	—	69 cm
Domestic apple *Malus domestica*	—	—	✓	—	—	✓	—	—	—	27 cm
London plane *Platanus × acerifolia*	—	—	—	—	—	—	—	—	✓	178 cm
Hybrid black poplar *Populus × canadensis*	✓	—	✓	—	—	—	—	—	—	117 cm

Disease & Decay PRF – *continued*

TREE SPECIES	DISEASE & DECAY PRF THE BROADLEAVED TREE SPECIES FORMS THAT ARE KNOWN TO BE EXPLOITED BY ROOSTING BATS (AS DEMONSTRATED BY RECORDS WITH ACCOMPANYING PHOTOGRAPHS ON THE BTHK DATABASE OR PHOTOGRAPHIC ACCOUNTS WITHIN BAT TREE HABITAT KEY (Andrews *et al.* 2016))									MINIMUM DBH OF DISEASE & DECAY ROOST TREE RECORDS FOR THE TREE SPECIES HELD ON DATABASE
	Woodpecker-holes	Squirrel-holes	Knot-holes	Pruning-cuts	Tear-outs	Wounds	Cankers	Compression-forks	Butt-rots	
Wild cherry *Prunus avium*	—	—	—	—	✓	—	✓	—	—	36 cm
Turkey oak *Quercus cerris*	—	—	—	—	—	—	—	—	✓	121 cm
Sessile oak *Quercus petraea*	✓	✓	✓	—	✓	✓	—	—	—	25 cm
Pedunculate oak *Quercus robur*	✓	—	✓	✓	✓	✓	—	—	—	13 cm
Goat/grey willow *Salix caprea/cinerea*	✓	—	—	—	✓	✓	—	—	—	19 cm
White/crack willow *Salix alba/fragilis*	✓	—	—	—	—	✓	—	—	—	54 cm
Rowan *Sorbus aucuparia*	—	—	—	—	—	✓	—	—	—	15 cm
Lime *Tilia* spp.	—	—	—	—	—	—	—	—	✓	258 cm
Elm *Ulmus* spp.	—	—	—	—	—	✓	—	—	—	7 cm

Damage PRF

TREE SPECIES	DAMAGE PRF THE BROADLEAVED TREE SPECIES FORMS THAT ARE KNOWN TO BE EXPLOITED BY ROOSTING BATS (AS DEMONSTRATED BY RECORDS WITH ACCOMPANYING PHOTOGRAPHS ON THE BTHK DATABASE OR PHOTOGRAPHIC ACCOUNTS WITHIN BAT TREE HABITAT KEY (Andrews *et al.* 2016))									MINIMUM DBH OF DAMAGE ROOST TREE RECORDS FOR THE TREE SPECIES HELD ON DATABASE
	Lightning-strikes	Hazard-beams	Subsidence-cracks	Shearing-cracks	Transverse-snaps	Welds	Lifting bark	Desiccation-fissures	Frost-cracks	
Horse chestnut *Aesculus hippocastanum*	—	✓	—	—	—	—	—	—	✓	57 cm
Downy birch *Betula pubescens*	—	—	✓	—	—	—	—	—	✓	12 cm
Sweet chestnut *Castanea sativa*	—	✓	—	—	—	—	✓	✓	—	60 cm
Beech *Fagus sylvatica*	✓	✓	✓	—	✓	✓	—	—	✓	15 cm
Ash *Fraxinus excelsior*	—	✓	✓	—	✓	—	✓	—	—	14 cm
Sessile oak *Quercus petraea*	✓	✓	✓	—	✓	✓	✓	—	✓	8 cm
Pedunculate oak *Quercus robur*	✓	✓	✓	✓	✓	—	✓	✓	—	22 cm
Black locust *Robinia pseudoacacia*	—	✓	—	—	—	—	—	—	—	Data deficient
Crack willow *Salix fragilis*	—	—	—	✓	—	—	—	—	—	231 cm
Lime *Tilia* spp.	—	—	—	—	—	✓	—	—	—	78 cm

Association PRF

TREE SPECIES	ASSOCIATION PRF THE BROADLEAVED TREE SPECIES FORMS THAT ARE KNOWN TO BE EXPLOITED BY ROOSTING BATS (AS DEMONSTRATED BY RECORDS WITH ACCOMPANYING PHOTOGRAPHS ON THE BTHK DATABASE OR PHOTOGRAPHIC ACCOUNTS WITHIN BAT TREE HABITAT KEY (Andrews *et al.* 2016))		MINIMUM DBH OF DISEASE & DECAY ROOST TREE RECORDS FOR THE TREE SPECIES HELD ON DATABASE
	Fluting	Ivy	
Downy birch *Betula pubescens*	✓	—	23 cm
Hornbeam *Carpinus betulus*	✓	—	157 cm
Beech *Fagus sylvatica*	✓	—	100 cm
Pedunculate oak *Quercus robur*	—	✓	78 cm
Black locust *Robinia pseudoacacia*	✓	—	Data deficient
Lime *Tilia* spp.	—	✓	148 cm

Coniferous tree species

Disease & Decay PRF

TREE SPECIES	DISEASE & DECAY PRF THE CONIFEROUS TREE SPECIES FORMS THAT ARE KNOWN TO BE EXPLOITED BY ROOSTING BATS (AS DEMONSTRATED BY RECORDS WITH ACCOMPANYING PHOTOGRAPHS ON THE BTHK DATABASE OR PHOTOGRAPHIC ACCOUNTS WITHIN BAT TREE HABITAT KEY (Andrews *et al.* 2016))									MINIMUM DBH OF DISEASE & DECAY ROOST TREE RECORDS FOR THE TREE SPECIES HELD ON DATABASE
	Woodpecker-holes	Squirrel-holes	Knot-holes	Pruning-cuts	Tear-outs	Wounds	Cankers	Compression-forks	Butt-rots	
European larch *Larix decidua*	—	—	—	—	—	—	—	✓	—	Data deficient
Corsican pine *Pinus nigra*	✓	—	—	—	—	—	—	✓	—	Data deficient
Scots pine *Pinus sylvatica*	✓	—	—	—	—	✓	—	✓	✓	50 cm
Douglas fir *Pseudotsuga menziesii*	—	—	—	—	—	—	—	✓	✓	Data deficient
Yew *Taxus baccata*	—	—	—	—	—	—	—	✓	—	47 cm

Damage PRF

TREE SPECIES	DAMAGE PRF THE CONIFEROUS TREE SPECIES FORMS THAT ARE KNOWN TO BE EXPLOITED BY ROOSTING BATS (AS DEMONSTRATED BY RECORDS WITH ACCOMPANYING PHOTOGRAPHS ON THE BTHK DATABASE OR PHOTOGRAPHIC ACCOUNTS WITHIN BAT TREE HABITAT KEY (Andrews *et al.* 2016))									MINIMUM DBH OF DAMAGE ROOST TREE RECORDS FOR THE TREE SPECIES HELD ON DATABASE
	Lightning-strikes	Hazard-beams	Subsidence-cracks	Shearing-cracks	Transverse-snaps	Welds	Lifting bark	Desiccation-fissures	Frost-cracks	
Monterey cypress *Cupressus macrocarpa*	—	✓	—	—	—	✓	—	—	—	73 cm
European larch *Larix decidua*	—	✓	—	—	—	—	—	—	—	0.9 cm
Scots pine *Pinus sylvestris*	—	✓	---	—-	—	—	—	—	—	Data deficient
Giant redwood (sequoia) *Sequoiadendron giganteum*	—	—	—	—-	—	—	✓	✓	—	189 cm
Yew *Taxus baccata*	—	✓	---	—-	—	—	—	—	—	58 cm

Association PRF

TREE SPECIES	ASSOCIATION PRF THE BROADLEAVED TREE SPECIES FORMS THAT ARE KNOWN TO BE EXPLOITED BY ROOSTING BATS (AS DEMONSTRATED BY RECORDS WITH ACCOMPANYING PHOTOGRAPHS ON THE BTHK DATABASE OR PHOTOGRAPHIC ACCOUNTS WITHIN BAT TREE HABITAT KEY (Andrews *et al.* 2016))		MINIMUM DBH OF DISEASE & DECAY ROOST TREE RECORDS FOR THE TREE SPECIES HELD ON DATABASE
	Fluting	Ivy	
Yew *Taxus baccata*	✓	—	43 cm

PRF Recording Form

GROUNDSMAN*			
SITE NAME			
DATE			
HABITAT (Phase 1 where known)			
OS GRID REFERENCE			
TAG NUMBER			
TREE SPECIES			
TREE ALIVE/DEAD			
DBH (Diameter at Breast Height)			
TREE HEIGHT			
PRF on STEM/LIMB			
PRF TYPE			
DIRECTION PRF FACES			

CLIMBER

MEASUREMENTS

PRF HEIGHT (m)		
DPH (Diameter at PRF Height)		
ENTRANCE	**Height (cm)**	
	Width (cm)	
INTERNAL	**Height (cm)**	
	Width (cm)	
	Depth (cm)	

ROOSTING EVIDENCE

Droppings: yes or no? (look in base)		
Bats (look in cavity)	**Species**	
	Number of bats	
	Torpid or awake	
	Above/below/to side of entrance	
	Distance from entrance (cm)	

INTERNAL CONDITIONS: (Tick as many of the following as are applicable. Where a field is negative please put a cross.) **There should be a value in EVERY box:** ✓ or ✕

SMELL	None	Pleasant	Not unpleasant	Unpleasant
SUBSTRATE	Smooth	Bobbly	Bumpy	Rough
	Clean =	Waxy	Blackened	Polished
	Dirty =	Dusty	Debris	Sludgy
HUMIDITY	Dry	Damp	Wet	Mildew
APEX SHAPE	Dome	Spire	Peak/wedge	Flat
	Chambered		Tube	
COMPETITORS (invertebrates, birds, mammals?)				

PHOTOS	1. Contextual photo showing tree in habitat	
	2. Close-up photo of the PRF	
FORM	Check the form to ensure every box has a value	

* It is the Groundsman's responsibility to ensure the form is correctly and comprehensively completed.

** Cleanliness categories have a primary (i.e. clean or dirty) and then any associated secondary is ticked where it is observed.

References

Andrews H and Gardener M 2016. *Bat Tree Habitat Key – Database Report 2016*. Bridgwater: AEcol.

Andrews H and Pearson L 2016. *A Review of Empirical Data in respect of Emergence and Return Times Reported for the UK's 17 Native Bat Species*. Bridgwater: AEcol.

Andrews H *et al.* 2016. *Bat Tree Habitat Key*, 3rd edn. Bridgwater: AEcol.

August T, Nunn M, Fensome A, Linton D and Mathers F 2014. Sympatric woodland *Myotis* bats form tight-knit groups with exclusive home ranges. *PLoS One* 9(10): e112225.

Avery M 1985. Winter activity in pipistrelle bats. *Journal of Animal Ecology* 54: 721–738.

Bat Conservation Trust 2010. *The National Bat Monitoring Programme: Annual Report 2009*. London: Bat Conservation Trust; Peterborough: Joint Nature Conservancy Council.

Bat Conservation Trust 2014. *The State of the UK's Bats 2014: National Bat Monitoring Programme Population Trends*. London: Bat Conservation Trust.

Berthinussen A and Altringham J 2012. Do bat gantries and underpasses help bats cross roads safely? *PLoS One* 7: e38775.

Billington G 2002. *Report on Further Research of Barbastelle Bats Associated with Pengelli Forest Special Area of Conservation*. Greena Ecological Consultancy report to Countryside Council for Wales.

Billington G 2004. *The Use of Tree Roosts by Barbastelle Bats*. Oral presentation at 'Managing Trees and Woodlands as Habitat for Bats' Conference, Bournemouth University.

Bright P, Morris P and Mitchell-Jones A 2006. *The Dormouse Conservation Handbook*, 2nd edn. Peterborough: English Nature.

British Standards Institute 2015. *BS 8956 – Surveying for Bats in Trees and Woodland*. London: BSI.

Chanin P and Woods M 2003. *Surveying Dormice Using Nest Tubes. Results and Experience from the South West Dormouse Project*. English Nature Research Report 524. Natural England.

CIEEM 2013. *Code of Professional Conduct*. Winchester: Chartered Institute of Ecology and Environmental Management.

CIEEM 2016. *Guidelines for Ecological Impact Assessment in the UK and Ireland: Terrestrial, Freshwater and Coastal*, 2nd edn. Winchester: Chartered Institute of Ecology and Environmental Management.

Collins J (ed.) 2016. *Bat Surveys for Professional Ecologists: Good Practice Guidelines*, 3rd edn. London: Bat Conservation Trust.

Craig P and de Búrca G 2015. *EU Law: Text, Cases, and Materials*. Oxford: Oxford University Press.

de Jong J 1995. Habitat use and species richness of bats in a patchy landscape. *Acta Theriologica* 40(3): 237–248.

Dense C and Rahmel U 2002. Untersuchungen zur Habitatnutzung der Großen Bartfledermaus (*Myotis brandtdii*) im nordwestlichen Niedersachsen. In A Meschede, K-G Heller and P Boye (eds), *Okolgie, Wanderungen und Genetik von Fledermausen in Waldern – Untersuchungen als Grundlage fur den Fledermausschutz*. Munster: Landsirtschaftsverlag. pp. 51–68. Cited in: Boye P and Dietz M 2005. *Development of Good Practice Guidelines for Woodland Managements for Bats*. English Nature Research Reports 661, Peterborough.

Dietz C and Kiefer A 2016. *Bats of Britain and Europe*. London: Bloomsbury.

Dietz C, Helversen O and Dietmar N 2011. *Bats of Britain, Europe and Northwest Africa*. London: A & C Black.

Dietz M and Pir J 2011. *Distribution, Ecology and Habitat Selection by Bechstein's Bat* (Myotis bechsteinii*) in Luxembourg*. Ökologie der Säugetiere 6. Laurenti Verlag.

Downs N and Racey P 2006. The use by bats of habitat features in mixed farmland in Scotland. *Acta Chiropterologica* 8(1): 169–185.

Downs N, Cresswell W, Reason P, Sutton G, Wells D and Wray S 2016. Sex-specific habitat preferences of foraging and commuting lesser horseshoe bats *Rhinolophus hipposideros* (Borkhausen, 1797) in lowland England. *Acta Chiropterologica* 18(2): 451–465.

Edlin H 1976. *Forestry Commission Booklet No. 15. Know Your Conifers*. London: HMSO.

Encarnação J, Kierdorf U, Holweg D, Janoch U and Wolters V 2005. Sex-related differences in roost-site selection by Daubenton's bats *Myotis daubentonii* during the nursery period. *Mammal Review* 35: 285–294.

Entwistle A, Racey P and Speakman J 1996. Habitat exploitation by a gleaning bat, *Plecotus auritus*. *Philosophical Transactions of the Royal Society of London Series B* 351(1342): 921–931.

European Commission 2007. *Guidance Document on the Strict Protection of Animal Species of Community Interest under the Habitats Directive 92/43/EEC*.

Ferguson-Lees J, Castell R and Leech D 2011. *A Field Guide to Monitoring Nests*. British Trust for Ornithology.

Feyerabend F and Simon M 2000. Use of roosts and roost switching in a summer colony of 45 kHz phonic type pipistrelle bats (*Pipistrellus pipistrellus* Schreber, 1774). *Myotis* 38: 51–59.

Frank R 1997. Zur Dynamik der Nutzung von Baumhöhlen durch ihre Erbauer und Folgenutzer am Beispiel des Philosophenwaldes in Gießen an der Lahn. Vogel und Umwelt. *Zeitschrift für Vogelkunde und Naturschutz in Hessen* 9: 59–84. Cited in: Boye P and Dietz M 2005 *Development of Good Practice Guidelines for Woodland Managements for Bats*. English Nature Research Reports 661, Peterborough.

Fuhrmann M 1991. Untersuchungen zur Biologie des Braunen Langohrs (*Plecotus auritus* L., 1758) im Lennebergwald bei Mainz. Diploma thesis, University of Mainz. Cited in: Boye P and Dietz M 2005 *Development of Good Practice Guidelines for Woodland Managements for Bats*. English Nature Research Reports 661, Peterborough.

Fuhrmann M and Seitz A 1992. Nocturnal activity of the brown long-eared bat (*Plecotus auritus* L. 1758): data from radio-tracking in the Lenneburg Forest near Mainz (Germany). In: I Priede and S Swift (eds), *Wildlife Telemetry: Remote Monitoring and Tracking of Animals*. Chichester: Ellis Horwood.

Fuhrmann M, Schreiber C and Tauchert J 2002. Telemetrische Untersuchungen an Bechstein-fledermäusen (*Myotis bechsteinii*) und Kleinen Abendseglern (*Nyctalus leisleri*) im Oberurseler Stadtwald und Umgebung (Hochtaunukreis). *Schriftenreihe für Landschartspflege und Naturschutz* 71: 131–140. Cited in: Dietz C, Helversen O and Dietmar N 2011. *Bats of Britain, Europe and Northwest Africa*. London: A & C Black.

Furmankiewicz J 2016. The social organization and behavior of the brown long-eared bat *Plecotus auritus*. In J Ortega (ed.), *Sociality in Bats*. Switzerland: Springer.

Geiger H and Rudolph B 2004. Wasserfledermaus, *Myotis daubentonii*. In A Meschede and B Rudolph (eds) *Fledermause in Bayern*. Ulmer Verlag. pp. 127–138. Cited in: Dietz C, Helversen O and Dietmar N 2011. *Bats of Britain, Europe and Northwest Africa*. London A & C Black.

Glover A and Altringham J 2008. Cave selection and use by swarming bat species. *Biological Conservation* 141: 1493–1504.

Green M 2013. *Night Vision*. http://www.visualexpert.com.

Greenaway F 2008. *Barbastelle Bats in the Sussex West Weald: 1997–2008*. West Weald Landscape Partnership, Sussex Wildlife Trust.

Harmer R, Kerr G and Thompson R 2010. *Managing Native Broadleaved Woodland*. London: TSO.

Harris S and Yalden D (eds) 2008. *Mammals of the British Isles: Handbook*, 4th edn. London: The Mammal Society.

Harris S, Morris P, Wray S and Yalden D 1995. *A Review of British Mammals: Population Estimates and Conservation Status of British Mammals other than Cetaceans*. Peterborough: Joint Nature Conservation Committee.

Heise G and Schmidt A 1988. Beiträge zur sozialen Organisation und Ökologie des Braunen Langohrs. *Nyctalus (NF)* 5: 445–465. Cited in: Boye P and Dietz M 2005 *Development of Good Practice Guidelines for Woodland Management for Bats*. English Nature Research Reports No. 661. Peterborough: Natural England.

Holdon P and Cleeves T 2002. *RSPB Handbook of British Birds*. London: A & C Black.

Hopkirk A and Russ J 2004. *Pre-Hibernal and Hibernal Activity and Dispersal Patterns of Leisler's Bat, Nyctalus leisleri, in Northern Ireland*. Report to Environment and Heritage Service, Northern Ireland.

Howe C 1997. An account of three tree roosts in Motherwell. *Scottish Bats* 4: 9–11.

Hutson A 1984. *Keds, Flat-Flies and Bat-Flies – Diptera, Hippoboscidae and Nycteribiidae*. Handbooks for the Identification British Insects, Vol. 1 0, Part 7. Royal Entomological Society of London.

Hutterer R, Ivanova T, Meyer-Cords C and Rodrigues L 2005. *Bat Migrations in Europe – A Review of Banding Data and Literature*. Naturschutz und Biologische Vielfalt Heft 28. Bonn: Federal Agency for Nature Conservation, 2005.

Jahelková H and Horáček I 2011. Mating system of a migratory bat, Nathusius' pipistrelle (*Pipistrellus nathusii*): different male strategies. *Acta Chiropterologica* 13(1): 123–137.

JNCC 2008. *UK Biodiversity Action Plan Priority Habitat Descriptions – Traditional Orchards*. Peterborough: Joint Nature Conservation Committee.

JNCC 2010. *Handbook for Phase 1 Habitat Survey – A Technique for Environmental Audit*. Joint Nature Conservation Committee, Peterborough: Joint Nature Conservation Committee.

Kerth C and Melber M 2009. Species-specific barrier effects of a motorway on the habitat use of two threatened forest-living bat species. *Biological Conservation* 142: 270–279.

Korsten E 2012. *Vleermuiskasten: Overzicht van toepassing, gebruik en succesfactoren*. Culemborg: Bureau Waardenburg BV.

Kunz T, Hodgkison R and Weise C 2009. *Methods of Capturing and Handling Bats*. In T Kunz and S Parsons (eds), *Ecological and Behavioral Methods for the Study of Bats*, 2nd edn. Baltimore, MD: John Hopkins University Press.

Lewis T and Taylor L 1964. Diurnal periodicity of flight by insects. *Transactions of the Royal Entomological Society of London* 116(15): 393–476.

Lindenmayer D, Margules C and Botkin D 2000. Indicators of biodiversity for ecologically sustainable forest management. *Conservation Biology* 14: 941–950.

Linton D 2009. Bat ecology and conservation in lowland farmland. DPhil thesis, Oxford University.

Lonsdale, D 1999. *Research for Amenity Trees No. 7: Principles of Tree Hazard Assessment and Management*. London: Forestry Commission.

Lučan R and Radil J 2010. Variability of foraging and roosting activities in adult females of Daubenton's bat (*Myotis daubentonii*) in different seasons. *Biologia* 65(6): 1072–1080.

Martin G 1990. *Birds by Night*. London: A & D Poyser.

Meschede A and Heller K 2000. *Ökologie und Schutz von Fledermäusen in Wäldern*. Schriftenreihe für Landschaftspflege und Naturschutz 66. Münster: Landwirtschaftsverlag. Cited in: Boye P and Dietz M 2005. *Development of Good Practice Guidelines for Woodland Managements for Bats*. English Nature Research Reports 661, Peterborough.

Mitchell-Jones A and McLeish A (eds) 2004. *The Bat Worker's Manual*, 3rd edn. Peterborough: Joint Nature Conservation Committee.

Murphy S, Greenaway F and Hill D 2012. Patterns of habitat use by female brown long-eared bats presage negative impacts of woodland conservation management. *Journal of Zoology*, doi:10.1111/j.1469–7998.2012.00936.x.

Natural England 2010a. *Traditional Orchards: Site and Tree Selection – Natural England Technical Information Note TIN013*. Peterborough: Natural England.

Natural England 2010b. *Traditional Orchards: Restoring and Managing Mature and Neglected Orchards – Natural England Technical Information Note TIN018*. Peterborough: Natural England.

Natural England 2013. *Guidance on the Capture and Marking of Bats under the Authority of a Natural England Licence: WML-G39 (10/13)*. Peterborough: Natural England.

Natural England 2016. *Proposed New Policies for European Protected Species Licensing: Analysis of Responses to the Public Consultation Held between 25 February and 7 April 2016–December 2016*. Peterborough: Natural England.

Palmer E, Pimley E, Sutton G and Birks J 2013. *A Study on the Population Size, Foraging Range and Roosting Ecology of Bechstein's bats at Grafton Wood SSSI, Worcestershire*. Report to the Peoples Trust for Endangered Species and Worcestershire Wildlife Trust by Link Ecology and Swift Ecology.

Parsons K, Jones G, Davidson-Watts I and Greenaway F 2003. Swarming of bats at underground sites in Britain – implications for conservation. *Biological Conservation* 111: 63–70.

Pénicaud P 2000. Chauves-souris arboricoles en Bretagne (France): typologie de 60 arbres-gîtes et éléments de l'écologie des espèces observes. *Le Rhinolophe* 14: 37–68.

Pollard E, Hooper M and Moore N 1974. *Hedges*. London: Collins New Naturalist Series.

Poulton S 2006. *An Analysis of the Use of Bat Boxes in England, Wales and Ireland for The Vincent Wildlife Trust*. County Galway, Ireland: Vincent Wildlife Trust.

Rackham O 1995. *Looking for Ancient Woodland in Ireland*. In J Pilcher and S Mac an tSaoir (eds), *Wood, Trees and Forests in Ireland*. Dublin: Royal Irish Academy. pp. 1 – 12.

Ransome R and Hutson A 2000. *Action Plan for the Conservation of the Greater Horseshoe Bat in Europe (*Rhinolophus ferrumequinum*)*. Strasbourg: Council of Europe.

Rodwell J (ed.) 1991. *British Plant Communities. Volume 1: Woodlands and Scrub*. Cambridge: Cambridge University Press.

Ruczyński I and Bogdanowicz W 2005. Roost cavity selection by *Nyctalus noctula* and *N. leisleri* (Vespertilionidae, Chiroptera) in Bialowieza primeval forest, Eastern Poland. *Journal of Mammalogy*, 86(5): 921–930.

Russ J 2012. *British Bat Calls – A Guide to Species Identification*. Exeter: Pelagic Publishing.

Russo D, Cisterone L and Jones G 2005. Spatial and temporal patterns of roost use by tree-dwelling barbastelle bats *Barbastella barbastellus*. *Ecography* 28: 769–776.

Sachanowicz K and Ruczyński I 2001. Summer roost sites of Myotis brandtii (Eveersmann, 1845) (Chiroptera, Vespertilionidae. in eastern Poland. *Mammalia* 65: 531–535.

Schaub A and Schnitzler H 2007. Flight and echolocation behaviour of three vespertilionid bat species while commuting on flyways. *Journal of Comparative Physiology A: Neuroethology, Sensory, Neural, and Behavioral Physiology* 193: 1185–1194.

Schorcht W 1998. Demökologische Untersuchungen am Kleinen Abendsegler *Nyctalus leisleri* (Kuhl 1817) in Südthüringen. Diploma thesis, University of Halle-Wittenberg. Cited in: Boye P and Dietz M 2005 *Development of Good Practice Guidelines for Woodland Managements for Bats*. English Nature Research Reports 661, Peterborough.

Sedgeley J and O'Donnell C 1999. Factors influencing the selection of roost cavities by a temperate rainforest bat (Vespertilionidae: *Chalinolobus tuberculatus*) in New Zealand. *Journal of Zoology (London)* 249: 437–446.

Shiel C and Fairley 1999. Evening emergence of two nursery colonies of Leisler's bat (*Nyctalus leisleri*) in Ireland. *Journal of Zoology (London)* 247: 439–447.

Simms E 1971. *Woodland Birds*. Collins New Naturalist Series No. 52. London: Collins.

Simon M, Hüttenbügel S and Smit-Viergutz J 2004. *Ecology and Conservation of Bats in Villages and Towns*. Bonn: Bundesamt für Naturschutz.

Smith P and Racey P 2008. Natterer's bats prefer foraging in broad-leaved woodlands and river corridors. *Journal of Zoology (London)* 275: 314–322.

Smith P 2001. Habitat preference, range use and roosting ecology of Natterer's bats (*Myotis nattereri*), in a grassland-woodland landscape. PhD thesis, University of Aberdeen. Cited in: Mackie I and Racey P (undated) *A Literature Review of the Use of Forests by Bats to Provide Information for Guidelines for the Management of Forests by Bats*. Report to the Bat Conservation Trust.

Sparks J and Soper T 1970. *Owls – Their Natural and Unnatural History*. Newton Abbot: David & Charles.

Stebbings R, Yalden D and Herman J 1986. *Which Bat Is It? A Guide to Bat Identification in Great Britain and Ireland*. London: The Mammal Society.

Stone E 2011. Bats and development: with a particular focus on the impacts of artificial lighting. PhD thesis, University of Bristol, Cited in: Stone E (ed.) 2013 *Bats and Lighting – Overview of Current Evidence and Mitigation*. Bristol: University of Bristol.

Stone E (ed.) 2013. *Bats and Lighting – Overview of Current Evidence and Mitigation*. Bristol: University of Bristol.

Stone E, Jones G, and Harris S 2012. Conserving energy at a cost to biodiversity? Impacts of LED lighting on bats. *Global Change Biology* 18(8) doi: 10.1111/j.1365-2486.2012.02705 x.

Thomson A and Rankin G 1923. *Britain's Birds and their Nests*. Edinburgh: W & R Chambers.

Verboom B and Huitema H 1997. The importance of linear-landscape-elements for the pipistrelle *Pipistrellus pipistrellus* and the serotine bat *Eptesicus serotinus*. *Landscape Ecology* 12(2): 117–125.

Watson R 2006. *Trees: Their Use, Management, Cultivation and Biology*. Marlborough: Crowood Press.

Whitaker A 1905. *Arthur Whitaker's Bats*. The Naturalist. Facsimile edition published by Sheffield Bat Group with kind permission of the Yorkshire Naturalist Union.

Zeale M 2011. Conservation biology of the barbastelle (*Barbastella barbastellus*) – applications of spatial modelling, ecology and molecular analysis of diet. PhD thesis, University of Bristol.

Zeale M, Davidson-Watts I and Jones G 2012. Home range use and habitat selection by barbastelle bats (*Barbastella barbastellus*): implications for conservation. *Journal of Mammalogy* 93(4): 1110–1118.

Bibliography

For those people who wish to deepen their knowledge of PRF formation and tree-roosting bat ecology, the following texts are recommended:

Tree-roost formation

Andrews *et al.* 2016. *Bat Tree Habitat Key*, 3rd edn. Bridgwater: AEcol. [Free download from www.battreehabitatkey.com.]

Lonsdale D 1999. *Research for Amenity Trees No. 7: Principles of Tree Hazard Assessment and Management*. London: Forestry Commission. [The master at work – a reference that will seldom spend time on the shelf.]

Lonsdale D 2000. *Hazards from Trees – A General Guide*. Edinburgh: Forestry Commission. [An outstanding resource by the master of PRF formation (whether he knows it or not), and free to download!]

Mattheck C and Breloer H 2010. *Research for Amenity Trees No. 4: The Body Language of Trees: A Handbook for Failure Analysis*. London: The Stationery Office. [A 'must-have' for anyone working with trees.]

Rackham O 1976. *Trees and Woodland in the British Landscape*. Archaeology in the Field Series. London: J. M. Dent & Sons Ltd. [Superb background reading.]

Rackham O 2003. *Ancient Woodland: Its History, Vegetation and Uses in England: New Edition*. Kirkudbrightshire: Castlepoint Press. [The alpha through omega of ancient woodland.]

Schwarze W, Engels J and Mattheck C 2000. *Fungal Strategies of Wood Decay in Trees*. Berlin: Springer. [Not essential unless you propose to really follow the rabbit-hole down.]

Shigo A 2008. *Modern Arboriculture: A Systems Approach to the Care of Trees and their Associates*. Snohomish, WA: Shigo and Trees Associates. [Not essential unless you propose to really follow the rabbit-hole down.]

Stokland J, Siitonen J and Jonsson B 2012. *Biodiversity in Dead Wood*. Cambridge: Cambridge University Press. [Not essential unless you propose to really follow the rabbit-hole down.]

Watson R 2006. *Trees: Their Use, Management, Cultivation and Biology*. Marlborough: The Crowood Press. [Still the most accessible and interesting read. Everyone working with trees should take a couple of days off every winter to re-read this.]

Tree-roosting ecology

Andrews *et al.* 2016. Bat Tree Habitat Key, 3rd edn. Bridgwater: AEcol. [Free download from www.battreehabitatkey.com.]

Andrews H and Gardener M 2016. *Bat Tree Habitat Key – Database Report 2016*. Bridgwater: AEcol. [Free download from www.battreehabitatkey.com.]

Dietz M and Pir J 2011. *Distribution, Ecology and Habitat Selection by Bechstein's bat (*Myotis bechsteinii*) in Luxembourg*. Ökologie der Säugetiere 6. [If only they would do all the other 12 tree-roosting species then I could go back to botany… I have bought this book three times and read each copy until it fell apart. As inspiring as it is interesting.]

Dietz M, Schieber K and Mehl-Rouschal C 2013. *Höhlenbäume im urbanum raum: Teil 2 – Leitfaden*. Frankfurt am Main: Der Magistrat der Stadt am Main, Umweltamt, Untere Naturschutzbehörde. [The German BTHK; a really insightful read with lots of photos. Thanks to the authors' generosity (lang lebe Einigkeit und Recht und Freiheit!) a free English translation can be downloaded from www.battreehabitatkey.com.]

Fitzsimons P, Hill D and Greenaway F 2002. *Patterns of Habitat Use by Female Bechstein's Bats (*Myotis bechsteinii*) from a Maternity Colony in a British Woodland.* School of Biological Sciences, University of Sussex. [Essential reading.]

Greenaway F 2008. *Barbastelle Bats in the Sussex West Weald: 1997–2008.* West Weald Landscape Partnership, Sussex Wildlife Trust. [Essential reading.]

Pénicaud P 2000. Chauves-souris arboricoles en Bretagne (France): typologie de 60 arbres-gîtes et éléments de l'écologie des espèces observes. *Le Rhinolophe* 14: 37–68. [My favourite work on the subject. The drawings offer a deeper insight. Everyone working with bats should read this and, thanks to the generosity of the author (vive la Francais!), a free English translation can be downloaded from www.battreehabitatkey.com.]

Zabel C and Anthony R 2003. *Mammal Community Dynamics – Management and Conservation of the Coniferous Forests of Western North America.* Cambridge: Cambridge University Press. [Fourth chapter only, which is 39 pages, so borrow if possible!]

Index

Page number in *italics* refer to figures, those in **bold** refer to tables and those <u>underlined</u> refer to forms and templates.